Table of Contents

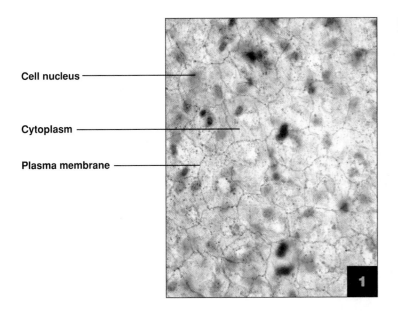

Cell nucleus

Cytoplasm

Plasma membrane

EPITHELIAL TISSUES

PLATE 1   Simple squamous epithelium, surface view. Silver stained mesothelium (360×)

DESCRIPTION: Single layer of flat cells with disc shaped central nuclei and little cytoplasm. In this surface view, the cells resemble fried eggs.

LOCATION: Form the kidney glomeruli and corpuscles; air sacs (alveoli) of the lungs; lining of the heart, blood vessels and lymphatic vessels; lining of the ventral body cavity (serosae).

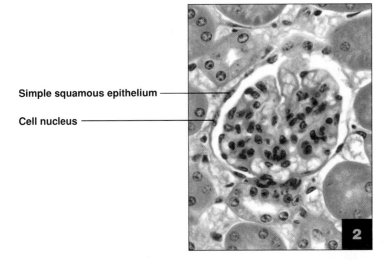

Simple squamous epithelium

Cell nucleus

PLATE 2   Simple squamous epithelium, section through a renal corpuscle in the renal cortex (465×)

DESCRIPTION: The single layer of flat cells with disc shaped central nuclei and small amount of cytoplasm are apparent in the parietal layer of the renal capsule. In this section, the thinness of the squamous cells and the dark staining central nuclei are obvious.

LOCATION: As listed for Plate 1.

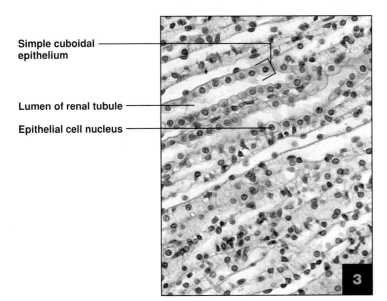

Simple cuboidal epithelium

Lumen of renal tubule

Epithelial cell nucleus

PLATE 3   Simple cuboidal epithelium, l.s. through renal medulla (350×)

DESCRIPTION: Single layer of cube-shaped cells with large round central nuclei. This section shows multiple rows of simple cuboidal epithelium forming the renal tubules.

LOCATION: Forms the kidney tubules and collecting ducts; the ducts and secretory portions of many glands; and the surface of the ovary.

From *A Brief Atlas of the Human Body*, Second Edition. Matt Hutchinson, Jon Mallatt, Elaine N. Marieb, and Patricia Brady Wilhelm. Copyright © 2007 by Pearson Education, Inc. Published by Pearson Benjamin Cummings. All rights reserved.

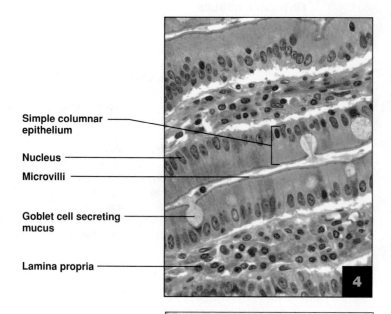

Simple columnar epithelium

Nucleus

Microvilli

Goblet cell secreting mucus

Lamina propria

4

PLATE 4   Non-ciliated simple columnar epithelium from the small intestine— jejunum (360X)

DESCRIPTION: Single layer of column shaped cells with either a round or oval shaped nucleus. Unicellular glands (goblet cells) that secrete mucous are common in this tissue. Microvilli, extensions of the plasma membrane of the apical surface, are present in the small intestine.

LOCATION: Lines digestive tract from stomach to anal canal; the gallbladder; portions of uterus and uterine tubes; and the excretory ducts of some glands.

Ciliated simple columnar epithelium

Nucleus

Lumen of uterine tube

Cilia

5

PLATE 5   Ciliated simple columnar epithelium from the oviduct (350×)

DESCRIPTION: Single layer of column shaped cells with either an oval or a round nucleus with cilia extending from the apical surface.

LOCATION: Lines the small bronchi, the uterine tubes, and portions of the uterus.

Pseudostratified columnar epithelium

Nuclei

Goblet cell

Cilia

6

PLATE 6   Pseudostratified columnar epithelium from the trachea (335×)

DESCRIPTION: Single layer of cells of differing heights, some not reaching the apical surface. Nuclei located at different levels give the appearance of a multilayered (stratified) tissue. Mucous secreting goblet cells are common in this tissue. There are both ciliated and non-ciliated types. Goblet cells and cilia are seen in this specimen.

LOCATION: The non-ciliated type lines the sperm carrying ducts (duct of epididymis, ductus deferens, ejaculatory duct, and mid-portion of the male urethra); the ciliated type lines the trachea and most of the upper respiratory tract.

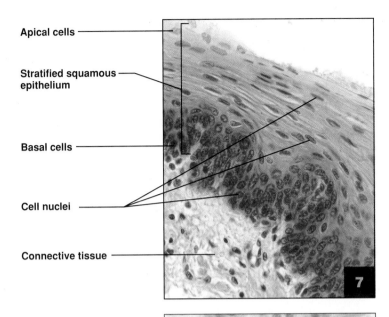

Apical cells

Stratified squamous epithelium

Basal cells

Cell nuclei

Connective tissue

7

PLATE 7   Stratified squamous epithelium from the mucosa of the esophagus (350×)

DESCRIPTION: Distinguished by multiple layers of cells with nuclei distributed throughout. The basal cells are cuboidal or columnar in shape, are metabolically active and have a high rate of mitosis; the apical cells are flat (squamous). In the keratinized type the surface cells are filled with the protein keratin.

LOCATION: The non-keratinized type (seen in this specimen) lines the esophagus, mouth, vagina and anus. The keratinized type forms the epidermis of the skin.

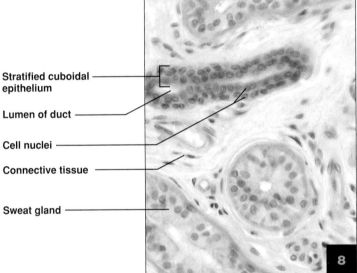

Stratified cuboidal epithelium

Lumen of duct

Cell nuclei

Connective tissue

Sweat gland

8

PLATE 8   Stratified cuboidal epithelium from the duct of a sweat gland (340×)

DESCRIPTION: Two layers of cells, identifiable by the two rows of round nuclei. The apical layer is composed of cube shaped cells.

LOCATION: Lines the largest ducts of sweat glands, the mammary glands, and the salivary glands.

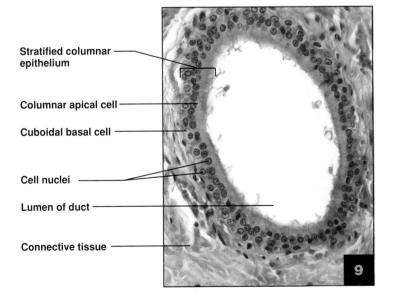

Stratified columnar epithelium

Columnar apical cell

Cuboidal basal cell

Cell nuclei

Lumen of duct

Connective tissue

9

PLATE 9   Stratified columnar epithelium from a duct in the parotid gland (350×)

DESCRIPTION: Two or more layers of cells: the apical layer is composed of columnar shaped cells, the basal layer is usually cuboidal cells.

LOCATION: Lines the large ducts of some glands, and some portions of the male urethra.

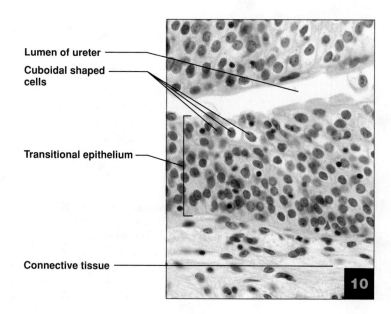

Lumen of ureter

Cuboidal shaped
cells

Transitional epithelium

Connective tissue

10

PLATE 10   Transitional epithelium from the
ureter (340×)

DESCRIPTION: As its name implies, this tissue
changes shape. It is composed of
multiple layers of cells. When in
relaxed state, as shown here, the
cells appear cuboidal in shape; when
stretched, the cells appear
squamous in shape.

LOCATION: Forms the lining of the urinary
bladder, ureter, and the superior
portion of the urethra.

Nuclei of fibroblasts

Collagen fiber

Elastic fibers

Gel-like ground substance

Mast cell

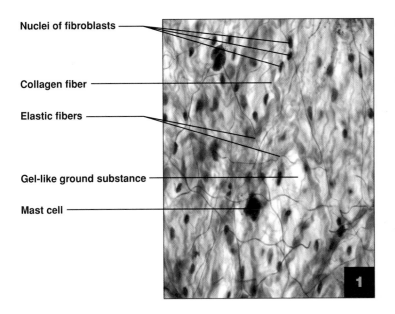

CONNECTIVE TISSUES—
CONNECTIVE TISSUE PROPER

PLATE 1   Areolar connective tissue (350×)

DESCRIPTION: Matrix contains all three fiber types (collagen fibers, elastic fibers, and reticular fibers) within a gel-like ground substance. Fibroblasts, mast cells, macrophages and other white blood cells are found within this tissue.

LOCATION: Distributed throughout the body loosely binding adjacent structures: forms the lamina propria that underlies all epithelia in the body; forms the papillary layer of the dermis of the skin and contributes to the superficial fascia; surrounds blood vessels, nerves, muscles.

PLATE 2   Adipose tissue from the external ear (350×)

DESCRIPTION: Distinguished by the closely packed adipocytes (fat cells) within a sparse matrix. Each adipocyte is filled with a large fat droplet causing the nucleus to be pushed to the edge of the cell.

LOCATION: A ubiquitous tissue found throughout the body: the hypodermis of the skin; surrounding the kidneys, eyeballs, mammary glands, and many other body organs; within the abdomen; and within the fascial planes separating muscle layers.

Cell nucleus of an adipocyte

Adipocytes

Fat droplet within an adipocyte

Blood vessel

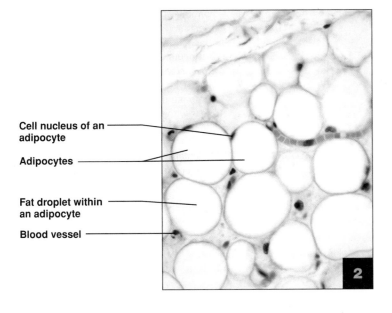

From *A Brief Atlas of the Human Body*, Second Edition. Matt Hutchinson, Jon Mallatt, Elaine N. Marieb, and Patricia Brady Wilhelm. Copyright © 2007 by Pearson Education, Inc. Published by Pearson Benjamin Cummings. All rights reserved.

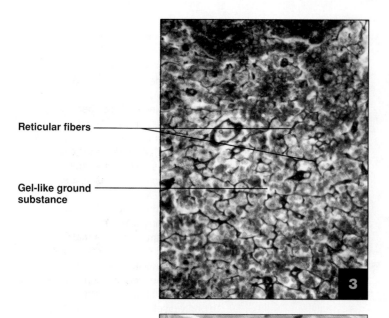

Reticular fibers

Gel-like ground substance

PLATE 3   Reticular connective tissue, lymph node (350×)

DESCRIPTION: Matrix is composed of reticular fibers loosely distributed within a gel-like ground substance. Cellular components are fibroblasts, lymphocytes, and other blood cells.

LOCATION: Forms the internal framework of many lymphoid organs: spleen, lymph nodes, bone marrow.

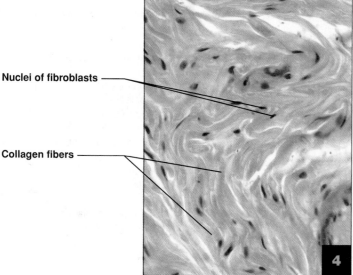

Nuclei of fibroblasts

Collagen fibers

PLATE 4   Dense irregular connective tissue from the submucosa of the large intestine (350×)

DESCRIPTION: Distinguished by the irregular arrangement of fibers densely packed in multiple directions. Composed primarily of collagen fibers with some elastic fibers. Major cell type is the fibroblast.

LOCATION: Reticular layer of dermis of skin, submucosa of digestive tract, fibrous capsules of organs and joints.

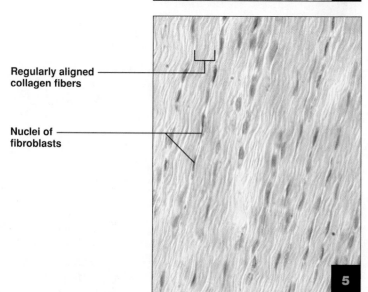

Regularly aligned collagen fibers

Nuclei of fibroblasts

PLATE 5   Dense regular connective tissue, tendon (340×)

DESCRIPTION: Densely packed fibers, primarily collagen, arranged parallel to each other. The nuclei of the fibroblasts are also aligned in parallel. This is an important feature for differentiating this tissue from smooth muscle tissue.

LOCATION: Tendons, most ligaments, aponeuroses.

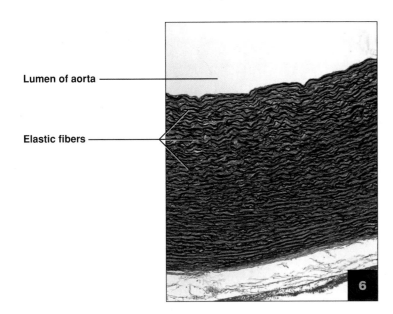

Lumen of aorta

Elastic fibers

6

PLATE 6   Elastic connective tissue from the aorta (90×)

DESCRIPTION: Connective tissue with densely packed elastic fibers. Notice the wavy appearance of the dark staining elastic fibers. Here, these fibers are within the smooth muscle layer of the wall of the aorta.

LOCATION: Found within the body wall of arteries; distributed throughout the trachea and bronchial tree; located within some tendons in the neck region.

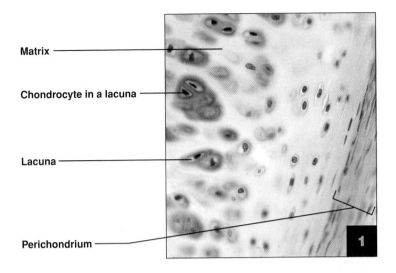

Matrix

Chondrocyte in a lacuna

Lacuna

Perichondrium

CONNECTIVE TISSUES—CARTILAGE

PLATE 1   Hyaline cartilage from the trachea (320×)

DESCRIPTION: Cartilage cells (chondrocytes) located within spaces (lacunae) in the tissue matrix. The matrix is a firm, gel-like ground substance embedded with collagen fibrils, which are not viewable via light microscopy. Vascularized perichondrium surrounds the cartilage nourishing the tissue and producing new tissue.

LOCATION: Forms most of the embryonic skeleton; covers the ends of long bones in joint cavities; forms the costal cartilages, the cartilages of the nose, trachea and bronchial tree, and most of the laryngeal cartilages.

PLATE 2   Elastic cartilage (350×)

DESCRIPTION: As in hyaline cartilage, the chondrocytes sit in spaces (lacunae) within the tissue matrix. The matrix contains a firm gel-like ground substance and both collagen fibrils and elastic fibers. In this preparation the elastic fibers are the dark purple strands viewable in the matrix.

LOCATION: Epiglottis and the external ear.

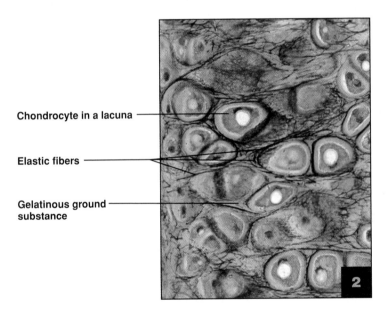

Chondrocyte in a lacuna

Elastic fibers

Gelatinous ground substance

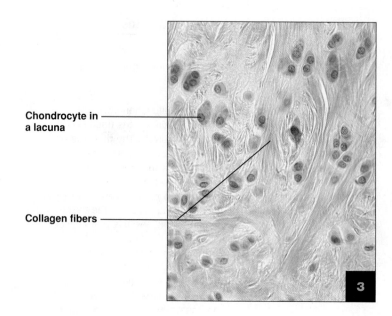

Chondrocyte in a lacuna

Collagen fibers

3

PLATE 3   Fibrocartilage, within a tendon (350×)

DESCRIPTION: The gelatinous matrix is densely packed with thick collagen fibers. Chondrocytes are located in lacunae, spaces in the matrix. This feature distinguishes this tissue from dense irregular connective tissue. This is the strongest type of cartilage.

LOCATION: Form the intervertebral discs; the pubic symphysis; the articular discs within joint cavities. Also located within some tendons, particularly where a tendon passes around a bony pulley.

Lamellae

Osteon

Central canal

Osteocytes in lacunae

Canaliculi extending
through matrix

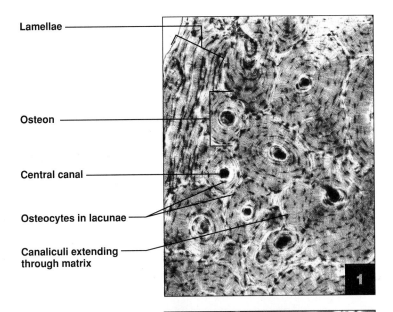

Red bone marrow

Trabecula

Osteocytes
within lacunae

Osteoblasts

Osteoclasts

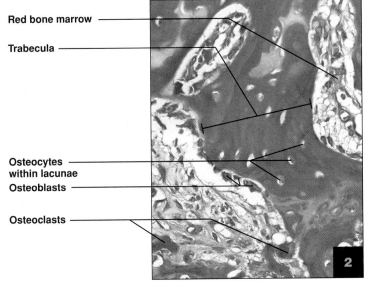

CONNECTIVE TISSUES—BONE

PLATE 1   Compact bone (90×)

DESCRIPTION: Tissue composed of a hard, calcified matrix containing many collagen fibers. This densely packed bone tissue is organized in lamellae (layers of bone tissue) and osteons (concentric rings of bone tissue). Blood vessels are located in the central canals; osteocytes lie in the lacunae; canaliculi, the thin dark lines, connect adjacent osteocytes.

LOCATION: Found in the shaft of long bones; the external portion of flat, short, and irregular shaped bones; and the external portion of the epiphyses.

PLATE 2   Spongy bone (340×)

DESCRIPTION: Composed of the same materials (hard calcified ground substance and collagen fibers) as compact bone. Arranged into small beams (trabeculae) of bony tissue. Spaces between trabeculae are filled with red bone marrow. Bone arranged in lamellae and osteocytes are located in lacunae. Bone forming cells, osteoblasts, line the trabeculae. Bone destroying cells, osteoclasts, also present as spongy bone continually remodels.

LOCATION: Located in the internal regions of the epiphyses, as well in the internal portions of flat, short, and irregular shaped bone.

Erythrocytes

Platelets

Leukocytes (neutrophils)

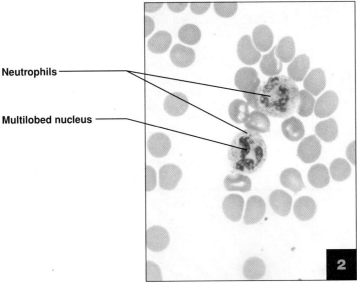
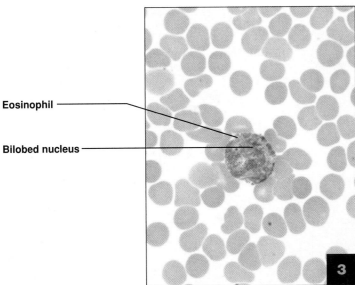

Neutrophils

Multilobed nucleus

Eosinophil

Bilobed nucleus

CONNECTIVE TISSUES—BLOOD

PLATE 1   Blood smear (270×)

DESCRIPTION: Erythrocytes (red blood cells), leukocytes (white blood cells), and platelets in a fluid matrix. Red blood cells, the small, red, disc-shaped cells shown here, are the most numerous formed element in blood.

LOCATION: Contained within blood vessels.

PLATE 2   Blood smear, neutrophils (895×)

DESCRIPTION: Granular leukocyte containing a multilobed nucleus. Cytoplasm appears light purple due to stain affinities. This is the most common leukocyte.

LOCATION: Originate and are stored in the red bone marrow. Travel in blood vessels. Enter loose connective tissues in response to infections. Respond to bacterial infections.

PLATE 3   Blood smear, eosinophil (885×)

DESCRIPTION: Granular leukocyte with a bilobed nucleus, cytoplasmic granules stain red. Relatively rare (1–4% of all leukocytes).

LOCATION: Originate and are stored in the red bone marrow. Travel in blood vessels. Enter loose connective tissues in response to infections. Respond to parasitic infection, and to end an allergic reaction.

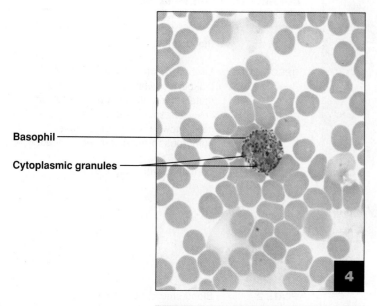

PLATE 4   Blood smear, basophil (840×)

DESCRIPTION: Granular leukocyte with a bilobed nucleus, cytoplasmic granules stain dark purple. Most rare (0–1% of all leukocytes).

LOCATION: Originate and are stored in the red bone marrow. Travel in blood vessels. Enter loose connective tissues in response to infections. Mediate inflammation by secreting histamines.

Basophil

Cytoplasmic granules

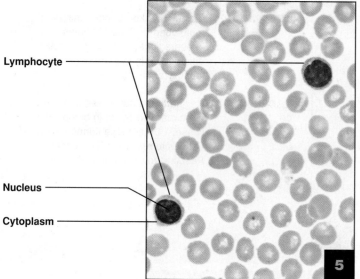

PLATE 5   Blood smear, lymphocytes (815×)

DESCRIPTION: Agranular leukocyte with a large circular nucleus that takes up most of the cell volume and stains purple, surrounded by a thin border of pale blue cytoplasm.

LOCATION: Originate in red bone marrow. Travel in blood vessels. Enter loose connective tissues and lymphoid tissue in response to infections.

Lymphocyte

Nucleus

Cytoplasm

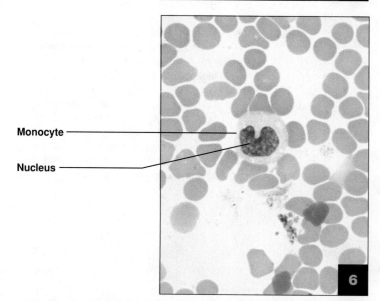

PLATE 6   Blood smear, monocyte (855×)

DESCRIPTION: Largest leukocyte; agranular with a kidney shaped nucleus that stains lighter than that of lymphocytes. Contains a larger amount of blue staining cytoplasm than in lymphocytes.

LOCATION: Originate and are stored in the red bone marrow. Travel in blood vessels. Enter loose connective tissues in response to infections, transform into macrophages to phagocytize foreign matter.

Monocyte

Nucleus

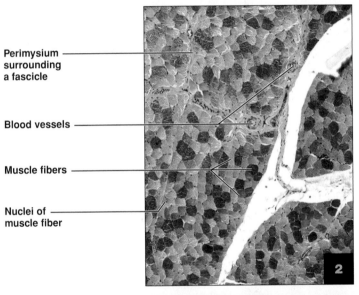

Muscle fiber

Nuclei

A band
I band
Z disc

Perimysium
surrounding
a fascicle

Blood vessels

Muscle fibers

Nuclei of
muscle fiber

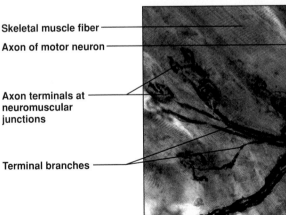

Skeletal muscle fiber
Axon of motor neuron

Axon terminals at
neuromuscular
junctions

Terminal branches

MUSCLE TISSUE—SKELETAL MUSCLE

PLATE 1   Skeletal muscle l.s. (445×)

DESCRIPTION: Long cylindrical cells (fibers), multinucleated, obvious striations running perpendicular to fiber direction. Dark bands are called A bands, light bands are I bands. In this section you can also see the Z discs, the thin dark lines running through the middle of the I bands.

LOCATION: In skeletal muscles.

PLATE 2   Skeletal muscle c.s. (85×)

DESCRIPTION: Cross section through skeletal muscle showing muscle fibers and connective tissues. Nuclei are located at the periphery of each fiber and the connective tissue, perimysium, (stained gray) groups bundles of muscle fibers into fascicles. Notice the blood vessels running within the perimysium.

LOCATION: In skeletal muscles.

PLATE 3   Neuromuscular junction (motor end plate; 290×)

DESCRIPTION: Junction of a motor neuron with skeletal muscle fibers. The axon branches into multiple axon terminals that innervate muscle fibers.

LOCATION: Within skeletal muscles.

MUSCLE TISSUE—CARDIAC MUSCLE

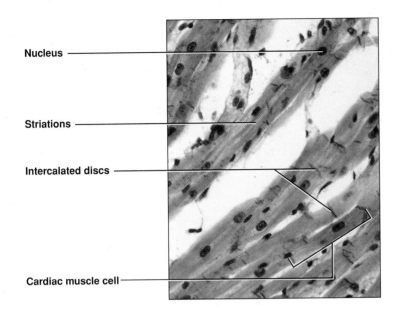

Nucleus

Striations

Intercalated discs

Cardiac muscle cell

Cardiac muscle (365×)

DESCRIPTION: Striated muscle (although striations are often difficult to view at this magnification) composed of branching cells with one centrally located nucleus (occasionally two). Cells are joined by specialized cell junctions, intercalated discs.

LOCATION: Makes up the myocardium of the heart.

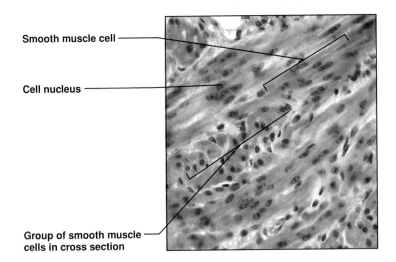

Smooth muscle cell ———

Cell nucleus ———

Group of smooth muscle
cells in cross section ———

MUSCLE TISSUE—SMOOTH MUSCLE

Smooth muscle, from the uterus
(300×)

DESCRIPTION: Non-striated muscle tissue.
Elongated, tapering cells with a
single, central nucleus are closely
packed together to form sheets.
Distinguishable from dense regular
connective tissue because nuclei are
randomly distributed throughout.

LOCATION: Composes the muscular layer in the
wall of the digestive tract, circulatory
vessels, urinary, and reproductive
organs. Also located in the
respiratory tubes and inside the eye.

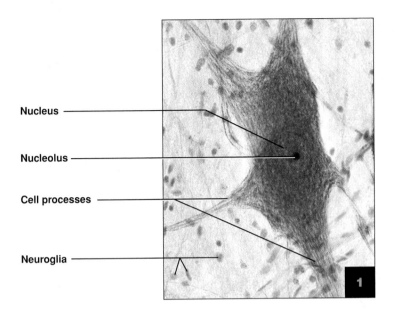

Nucleus

Nucleolus

Cell processes

Neuroglia

NERVOUS TISSUE

PLATE 1   Neuronal cell body in central nervous system (350×)

DESCRIPTION: The neuronal cell body contains the nucleus and cellular organelles. Extending from the cell body are cell processes that either receive or transmit signals. The dark staining fibers are neurofibrils, intermediate filaments that run throughout the neuron. This image also shows multiple neuroglia, supportive cells that aid neuronal function.

LOCATION: Found in the central nervous system in the brain and spinal cord; in the peripheral nervous system in peripheral ganglia.

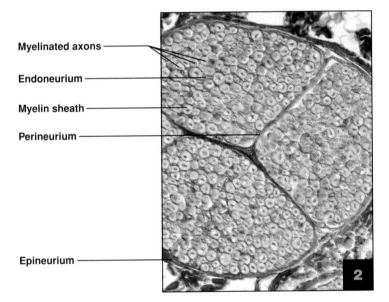

Myelinated axons

Endoneurium

Myelin sheath

Perineurium

Epineurium

PLATE 2   Peripheral nerve, c.s. (260×)

DESCRIPTION: A nerve is composed of the axonal processes of numerous neurons. Myelin surrounds many of the axons. Connective tissues wrap the axons: the endoneurium surrounds each axon; the perineurium bundles groups of axons into fascicles; the epineurium covers the entire nerve.

LOCATION: Found throughout the body carrying sensory and motor innervation from/to the periphery.

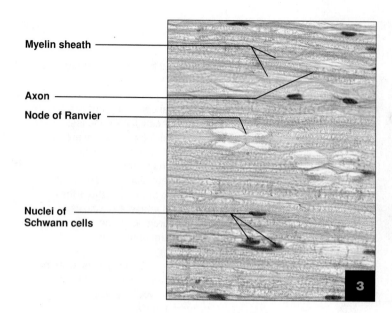

Myelin sheath

Axon

Node of Ranvier

Nuclei of
Schwann cells

3

PLATE 3   Peripheral nerve, l.s. (830×)

DESCRIPTION: Longitudinal section through a nerve
showing multiple axons surrounded
by myelin sheaths; Nodes of Ranvier,
pinching of the myelin sheath that
indicates the boundary between
adjacent Schwann cells; and
Schwann cell nuclei.

LOCATION: Found throughout the body
innervating body organs.

SELECT ORGANS

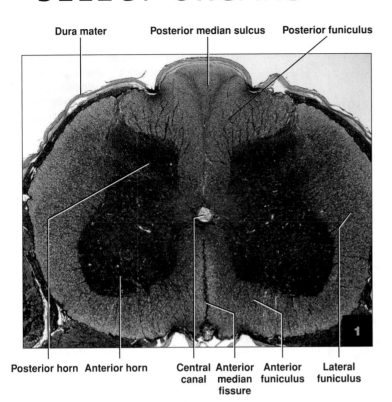

Dura mater — **Posterior median sulcus** — **Posterior funiculus**

Posterior horn — **Anterior horn** — **Central canal** — **Anterior median fissure** — **Anterior funiculus** — **Lateral funiculus**

PLATE 1

Spinal cord c.s. through lumbar region (18×)

DESCRIPTION: The hollow central canal is surrounded by a butterfly-shaped region of gray matter forming the anterior and posterior horns (stained brown in this section) that contain neuronal cell bodies; short, non-myelinated interneurons; and neuroglia. The external white matter, the funiculi, (stained gray) contains myelinated axons that make up the ascending and descending spinal tracts.

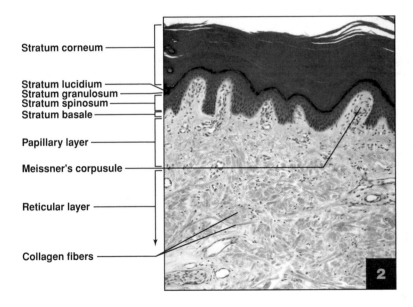

Stratum corneum

Stratum lucidium
Stratum granulosum
Stratum spinosum
Stratum basale

Papillary layer

Meissner's corpusule

Reticular layer

Collagen fibers

PLATE 2

Thick skin showing epidermal and dermal regions (85×)

DESCRIPTION: The epidermis, the dark pink region, is composed of the stratum basale, stratum spinosum, stratum granulosum, stratum lucidium, and the thick stratum corneum. Deep to the epidermis is the papillary layer of the dermis, composed of loose areolar connective tissue. Note the Meissner's corpuscle, a touch receptor, in the dermal papilla. The deepest layer of the dermis, the reticular layer, is composed of dense irregular connective tissue. Note the greater density of collagen fibers (stained pink) in this layer.

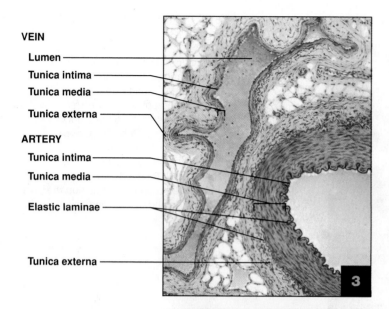

VEIN

Lumen

Tunica intima

Tunica media

Tunica externa

ARTERY

Tunica intima

Tunica media

Elastic laminae

Tunica externa

PLATE 3   Muscular artery and vein (80×)

DESCRIPTION: The vessel on the right, the artery, shows a wavy tunica intima resulting from the internal elastic lamina just deep to this layer. The thick tunica media is composed of multiple cell layers. Another layer of wavy elastic tissue, the external elastic lamina, is located outside of the tunica media. In the vein on the left, the lumen is irregular in shape, there are no elastic laminae, and the tunica media is only a few cells in thickness.

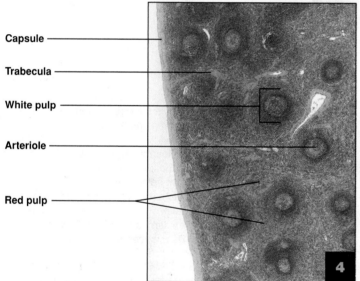

Capsule

Trabecula

White pulp

Arteriole

Red pulp

PLATE 4   Spleen (17×)

DESCRIPTION: The outer capsule of the spleen is reticular connective tissue. Trabeculae, extensions of this tissue into the deeper portion of the spleen, provide a framework for the organ. Lymphoid tissue, containing B and T lymphocytes surrounding arterial branches forms the white pulp of the spleen. Surrounding these "islands" is the red pulp, composed of both venous sinuses and splenic cords.

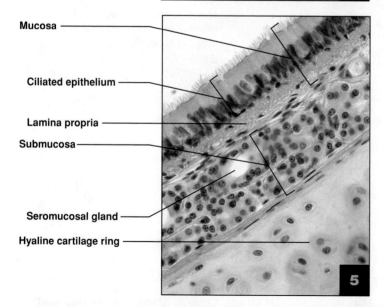

Mucosa

Ciliated epithelium

Lamina propria

Submucosa

Seromucosal gland

Hyaline cartilage ring

PLATE 5   Trachea (355×)

DESCRIPTION: The mucosa, composed of pseudostratified ciliated columnar epithelium and the underlying lamina propria, lines the trachea. Note the columnar epithelial cells, multiple nuclei, and the distinctive cilia along the apical surface. External to this layer, the submucosa consists of connective tissue and imbedded seromucosal glands that secrete mucous. Cartilagenous rings, composed of hyaline cartilage make up the most external layer seen.

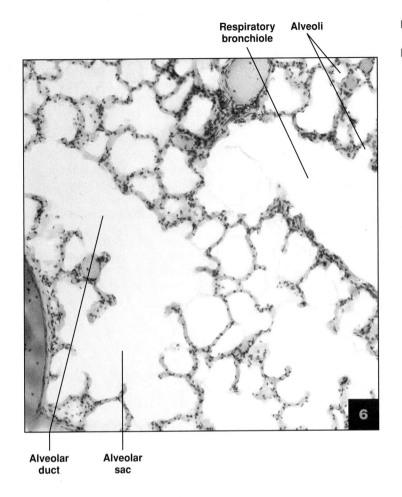

Respiratory bronchiole

Alveoli

Alveolar duct

Alveolar sac

PLATE 6   Lung (120×)

DESCRIPTION: This section through the lung shows the structures of the respiratory zone. The respiratory bronchiole, lined with simple cuboidal epithelium, has alveoli outpocketing from its wall. Alveolar ducts lead to alveolar sacs, terminal clusters of alveoli. Simple squamous epithelium forms the alveolar walls.

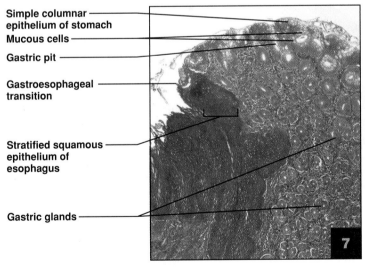

Simple columnar epithelium of stomach

Mucous cells

Gastric pit

Gastroesophageal transition

Stratified squamous epithelium of esophagus

Gastric glands

PLATE 7   Gastroesophageal junction (120×)

DESCRIPTION: The epithelial tissue changes abruptly at the gastroesophageal junction, from stratified squamous epithelium in the esophagus (on the left side of the image) to simple columnar epithelium of the stomach (top of the image). The gastric pits and gastric glands of the mucosal layer of the stomach, also composed of simple columnar epithelium, are apparent.

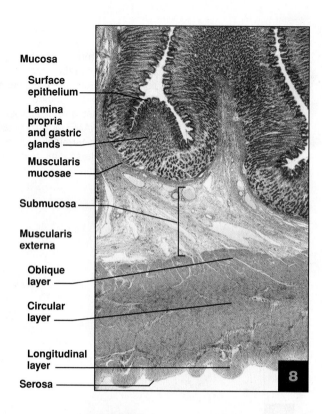

Mucosa
Surface epithelium
Lamina propria and gastric glands
Muscularis mucosae
Submucosa
Muscularis externa
Oblique layer
Circular layer
Longitudinal layer
Serosa

8

PLATE 8   Stomach, l.s. (17×)

DESCRIPTION: The mucosa is composed of the simple columnar epithelium forming the surface epithelium and the gastric glands, the lamina propria surrounding the glandular epithelium, and the muscularis mucosae, a thin layer of smooth muscle. The submucosa is a moderately dense connective tissue. The muscularis externa is three layers of smooth muscle: a deep oblique layer (difficult to distinguish at this magnification), a middle circular layer, and an external longitudinal layer. The serosa, the outermost layer, is loose areolar connective tissue and mesothelium.

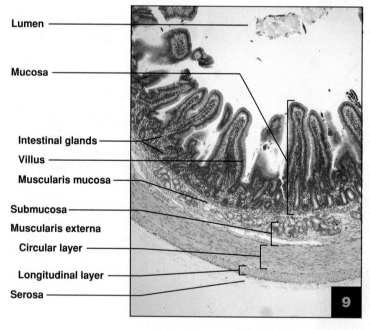

Lumen
Mucosa
Intestinal glands
Villus
Muscularis mucosa
Submucosa
Muscularis externa
Circular layer
Longitudinal layer
Serosa

9

PLATE 9   Small intestine, c.s. through duodenum (37×)

DESCRIPTION: The four layers of the wall of the small intestine are shown: the innermost layer, the mucosa, showing villi extending into the lumen, intestinal glands, and the muscularis mucosae; the submucosa, connective tissue imbedded with duodenal glands (only in the duodenum); the two layers of the muscularis externa, inner circular and outer longitudinal layers; and the outermost serosa composed of loose areolar connective tissue and simple squamous epithelium (mesothelium).

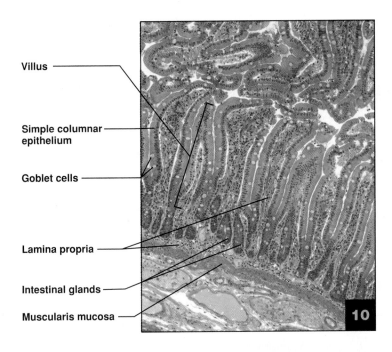

Villus

Simple columnar epithelium

Goblet cells

Lamina propria

Intestinal glands

Muscularis mucosa

10

PLATE 10 Mucosal layer of small intestine, from the jejunum (76×)

DESCRIPTION: Details of the mucosal layer are shown. Villi, lined with simple columnar epithelium and goblet cells, extend into the intestinal lumen; intestinal glands at the base of the villi produce digestive secretions; muscularis mucosae, a smooth muscle layer, forms the outermost layer of the mucosa. Refer to Plate 4 for high power view of mucosa. View the microvilli that make up the brush border on the apical surface of the columnar epithelial cells.

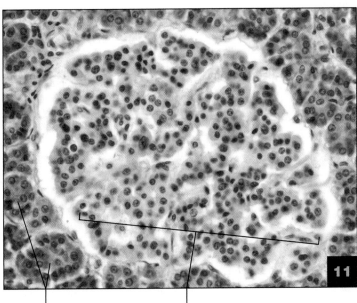

Exocrine pancreas, acinar cells

Endocrine pancreas, pancreatic islet

11

PLATE 11 Pancreas (350×)

DESCRIPTION: The glandular cells in the center of the field make up the pancreatic islet, the endocrine portion of the pancreas that produces insulin (beta cells), glucagon (alpha cells), and somatostatin (delta cells). These hormones are secreted into capillaries surrounding these cells. The exocrine pancreas, the acinar cells surround the islet. These cell produce digestive secretions that empty into the duodenum via ducts.

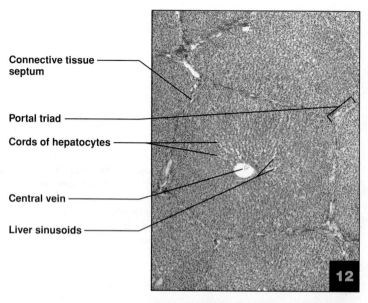

Connective tissue
septum

Portal triad

Cords of hepatocytes

Central vein

Liver sinusoids

PLATE 12   Pig liver (34×)

DESCRIPTION: Liver lobules are not distinct in
humans. Pig liver is shown here to
illustrate the lobular structure. A
central vein runs through the middle
of the lobule. Radiating outward are
cords containing hepatocytes. Blood
sinusoids are located between the
hepatic cords. Hepatic triads,
composed of branches of the
hepatic artery, hepatic portal vein,
and the bile duct, are found at the
corners of the lobule. Individual
lobules are separated by connective
tissue septa.

Renal corpuscle

Capsular space

Glomerular capillaries

Parietal layer of
glomerular capsule

Juxtaglomerular apparatus

Juxtaglomerular cells

Macula densa

Proximal convoluted
tubules

Distal convoluted tubule

PLATE 13   Renal cortex (270×)

DESCRIPTION: Two renal corpuscles are shown
surrounded by renal tubules in cross
section. Simple squamous
epithelium makes up the outer
portion of the glomerular capsule
and simple cuboidal epithelium
forms the portions of the renal
tubules shown. The proximal
convoluted tubules appear thicker
due to the microvilli extending from
their apical surface. The corpuscle
on the right shows the elongated
macula densa cells and
juxtaglomerular cells of the
juxtaglomerular apparatus.

Collecting duct

Thin segment of
loop of Henle

PLATE 14   Renal medulla (350×)

DESCRIPTION: Longitudinal section through renal
tubules in the renal medulla. Simple
cuboidal epithelium lines the
collecting ducts and the thick
segments of the loop of Henle (part
of the ascending limb). Simple
squamous epithelium forms the thin
segments of the loop of Henle.

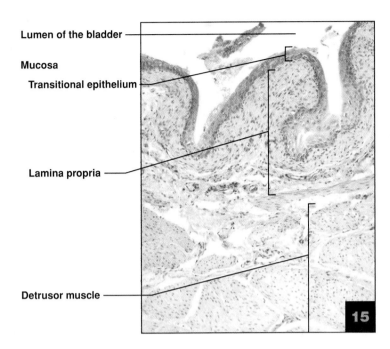

Lumen of the bladder

Mucosa

Transitional epithelium

Lamina propria

Detrusor muscle

15

PLATE 15   Urinary bladder (97×)

DESCRIPTION: The transitional epithelium lining the urinary bladder and the lamina propria, composed of loose areolar connective tissue, together make up the mucosa of the bladder. The thick smooth muscle layer, the detrusor muscle, is made up of three indistinct layers. The outer layer, the adventitia, is not shown here.

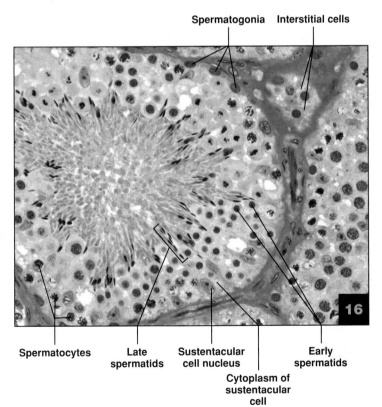

Spermatogonia   Interstitial cells

Spermatocytes   Late spermatids   Sustentacular cell nucleus   Early spermatids

Cytoplasm of sustentacular cell

16

PLATE 16   Testes (340×)

DESCRIPTION: This cross section through a seminiferous tubule shows spermatogenic cells embedded within columnar sustentacular cells. Stem cells, spermatogonia, are located at the periphery of the tubule. As cells move through the tubule toward the lumen, sperm are formed. Clusters of cells located between the tubules, interstitial cells, secrete testosterone.

Secondary follicle **Primordial follicles**

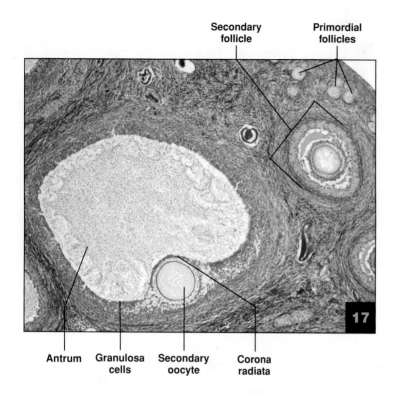

17

Antrum **Granulosa cells** **Secondary oocyte** **Corona radiata**

PLATE 17 Ovary (82×, trichrome stain)

DESCRIPTION: This section through an ovary shows follicles in various stages of development. The large follicle is a mature vesicular (Graafian) follicle. It contains a large fluid filled space, the antrum, lined by granulosa cells. Surrounding the oocyte is a ring of granulosa cells called the corona radiata. On the right side of the image, a secondary follicle is shown with the beginning of the antrum forming. The upper right corner contains multiple primordial follicles.

Endomytrium

Stratum functionalis

Coiled uterine glands

Stratum basalis

Myometrium

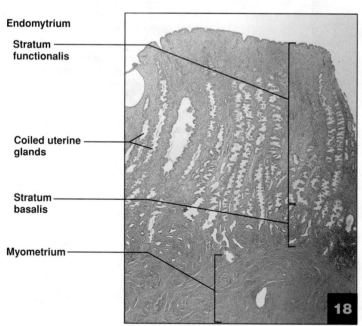

18

PLATE 18 Uterus, secretory stage (15×)

DESCRIPTION: The lumen of the uterus is at the top of the image. The thick stratum functionalis of the endometrium shows enlarged, coiled uterine glands. The thin stratum basalis forms the base of the endometrium. The myometrium, the smooth muscle layer of the uterine wall, is at the bottom of the image.

Colloid filled follicles

Follicle cells

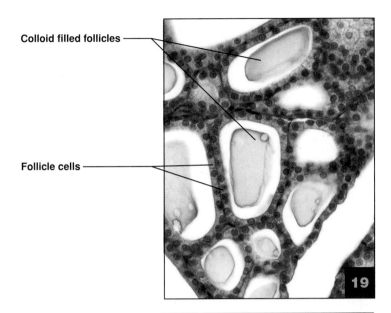

PLATE 19   Thyroid gland (350×)

DESCRIPTION: The thyroid gland is composed of spherical follicles formed by simple cuboidal epithelial cells, follicle cells. The center of each follicle is filled with a gel-like substance called colloid that contains proteins needed for the formation of thyroid hormone. Thyroid hormone is produced by the follicle cells and secreted into the capillaries that surround the follicles. In this image the colloid has pulled away from the follicle cells, an artifact of slide preparation.

Capsule

Zona glomerulosa

Zona fasciculata

Zona reticularis

Medulla

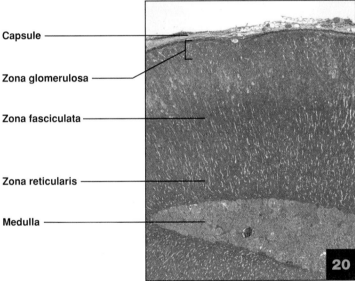

PLATE 20   Adrenal gland, section (35×)

DESCRIPTION: This section through the adrenal gland shows the medulla, light pink oval in the bottom of the image, whose cells secrete epinephrine and norepinephrine; the portions of the cortex: the cells of the thin zona reticularis and the thicker zona fasciculata secrete glucocortidoid hormones, and the cells of the superficial zona glomerulosa secrete mineralocorticoid hormones. The connective tissue forming the adrenal capsule is at the top of the image.

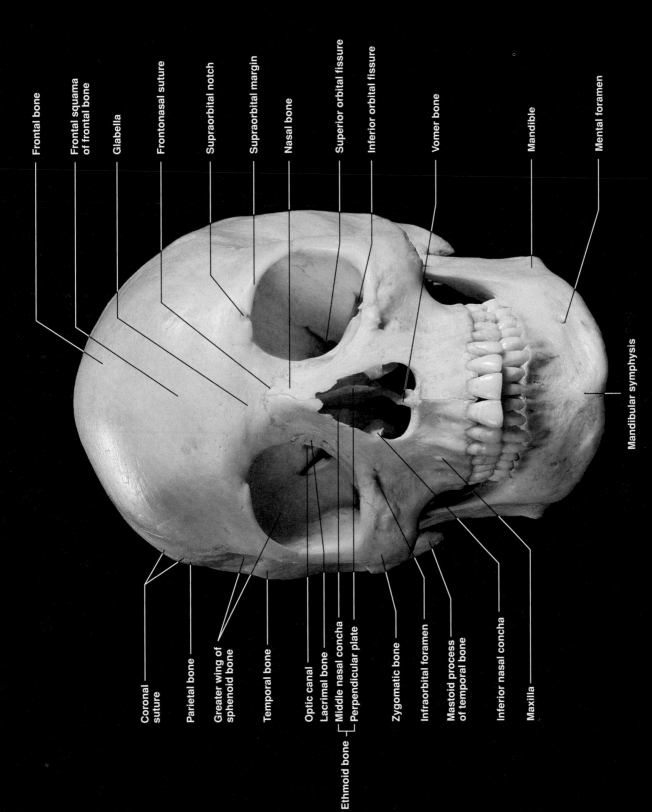

Skull, anterior view.

Frontal bone

Frontal squama of frontal bone

Glabella

Frontonasal suture

Supraorbital notch

Supraorbital margin

Nasal bone

Superior orbital fissure

Inferior orbital fissure

Vomer bone

Mandible

Mental foramen

Coronal suture

Parietal bone

Greater wing of sphenoid bone

Temporal bone

Optic canal

Lacrimal bone

Ethmoid bone — Middle nasal concha

Perpendicular plate

Zygomatic bone

Infraorbital foramen

Mastoid process of temporal bone

Inferior nasal concha

Maxilla

Mandibular symphysis

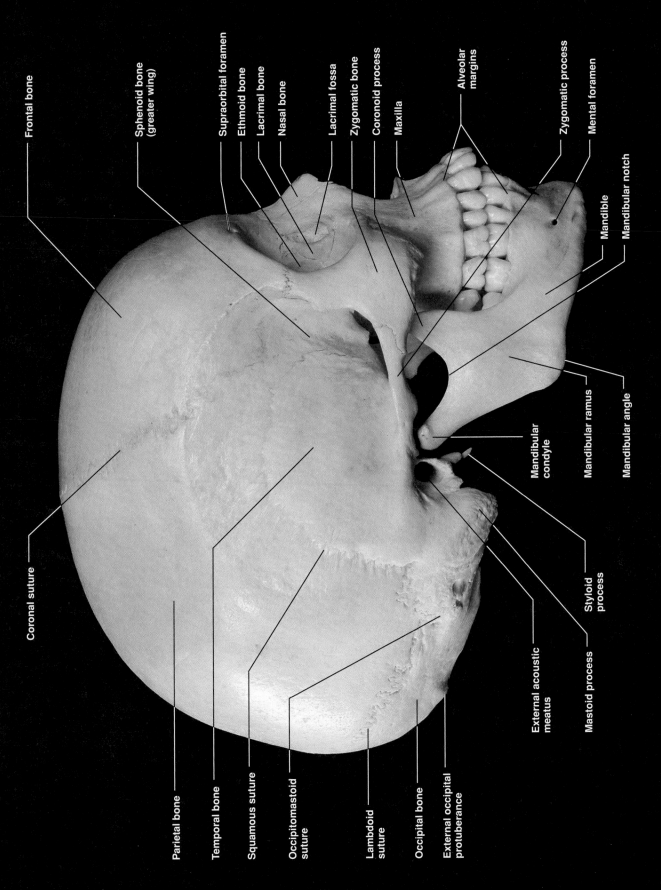

Frontal bone

Sphenoid bone (greater wing)

Supraorbital foramen

Ethmoid bone

Lacrimal bone

Nasal bone

Lacrimal fossa

Zygomatic bone

Coronoid process

Maxilla

Alveolar margins

Zygomatic process

Mental foramen

Mandible

Mandibular notch

Coronal suture

Parietal bone

Temporal bone

Squamous suture

Occipitomastoid suture

Lambdoid suture

Occipital bone

External occipital protuberance

External acoustic meatus

Mastoid process

Styloid process

Mandibular condyle

Mandibular ramus

Mandibular angle

Skull, right external view of lateral surface.

From *A Brief Atlas of the Human Body*, Second Edition. Matt Hutchinson, Jon Mallatt, Elaine N. Marieb, and Patricia Brady Wilhelm. Copyright © 2007 by Pearson Education, Inc. Published by Pearson Benjamin Cummings. All rights reserved.

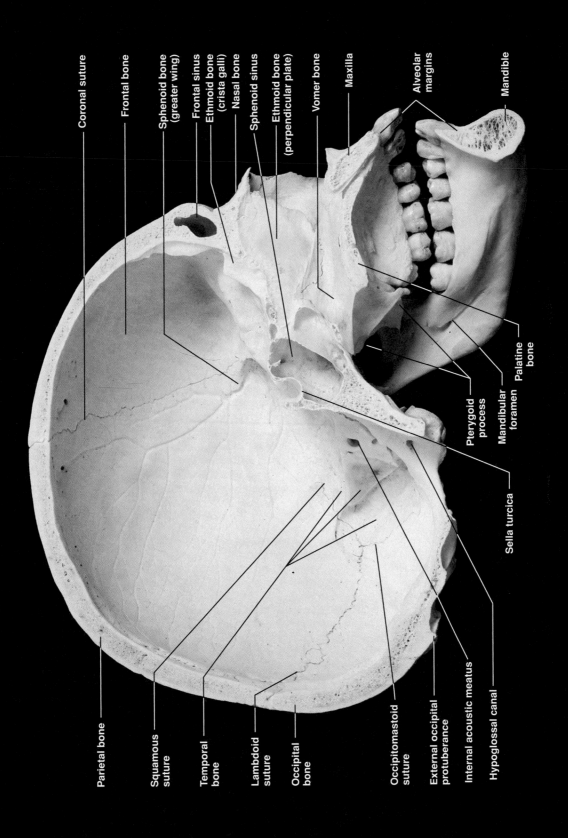

Coronal suture

Frontal bone

Sphenoid bone
(greater wing)

Frontal sinus

Ethmoid bone
(crista galli)

Nasal bone

Sphenoid sinus

Ethmoid bone
(perpendicular plate)

Vomer bone

Maxilla

Alveolar
margins

Mandible

Parietal bone

Squamous
suture

Temporal
bone

Lambdoid
suture

Occipital
bone

Occipitomastoid
suture

External occipital
protuberance

Internal acoustic meatus

Hypoglossal canal

Sella turcica

Pterygoid
process

Mandibular
foramen

Palatine
bone

Skull, internal view of left lateral aspect.

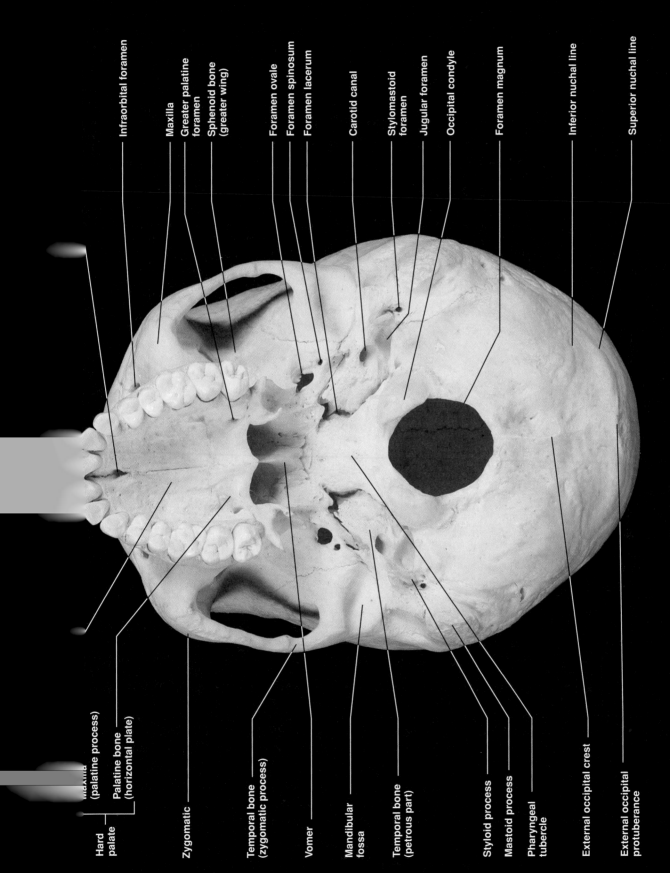

Infraorbital foramen

Maxilla

Greater palatine
foramen

Sphenoid bone
(greater wing)

Foramen ovale

Foramen spinosum

Foramen lacerum

Carotid canal

Stylomastoid
foramen

Jugular foramen

Occipital condyle

Foramen magnum

Inferior nuchal line

Superior nuchal line

Maxilla
(palatine process)

Palatine bone
(horizontal plate)

Hard
palate

Zygomatic

Temporal bone
(zygomatic process)

Vomer

Mandibular
fossa

Temporal bone
(petrous part)

Styloid process

Mastoid process

Pharyngeal
tubercle

External occipital crest

External occipital
protuberance

Skull, external view of base.

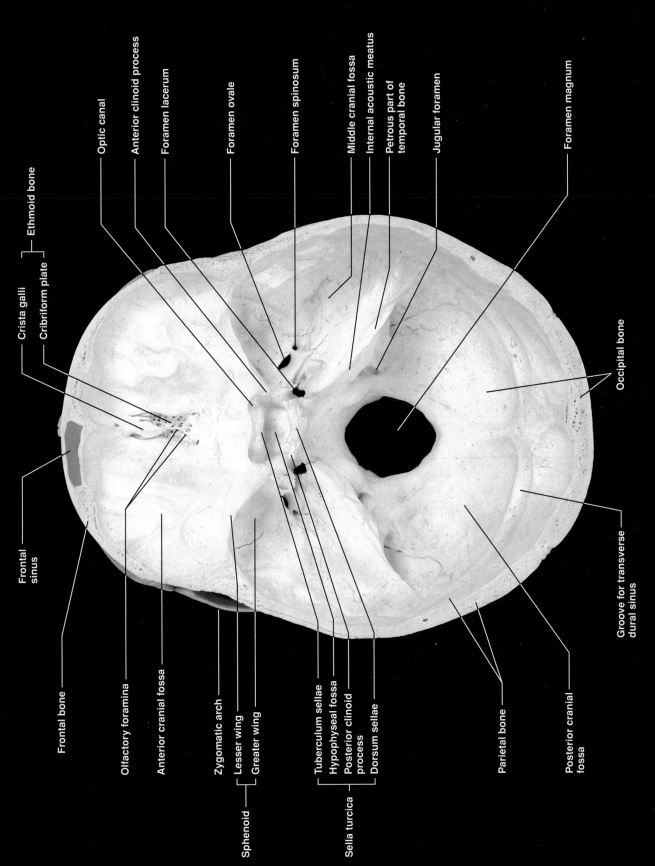

Crista galli — Ethmoid bone
Cribriform plate

Optic canal

Anterior clinoid process

Foramen lacerum

Foramen ovale

Foramen spinosum

Middle cranial fossa

Internal acoustic meatus

Petrous part of temporal bone

Jugular foramen

Foramen magnum

Occipital bone

Frontal sinus

Frontal bone

Olfactory foramina

Anterior cranial fossa

Zygomatic arch

Sphenoid — Lesser wing
Greater wing

Tuberculum sellae

Sella turcica — Hypophyseal fossa
Posterior clinoid process
Dorsum sellae

Parietal bone

Posterior cranial fossa

Groove for transverse dural sinus

Skull, internal view of base.

From *A Brief Atlas of the Human Body*, Second Edition. Matt Hutchinson, Jon Mallatt, Elaine N. Marieb, and Patricia Brady Wilhelm. Copyright © 2007 by Pearson Education, Inc. Published by Pearson Benjamin Cummings. All rights reserved.

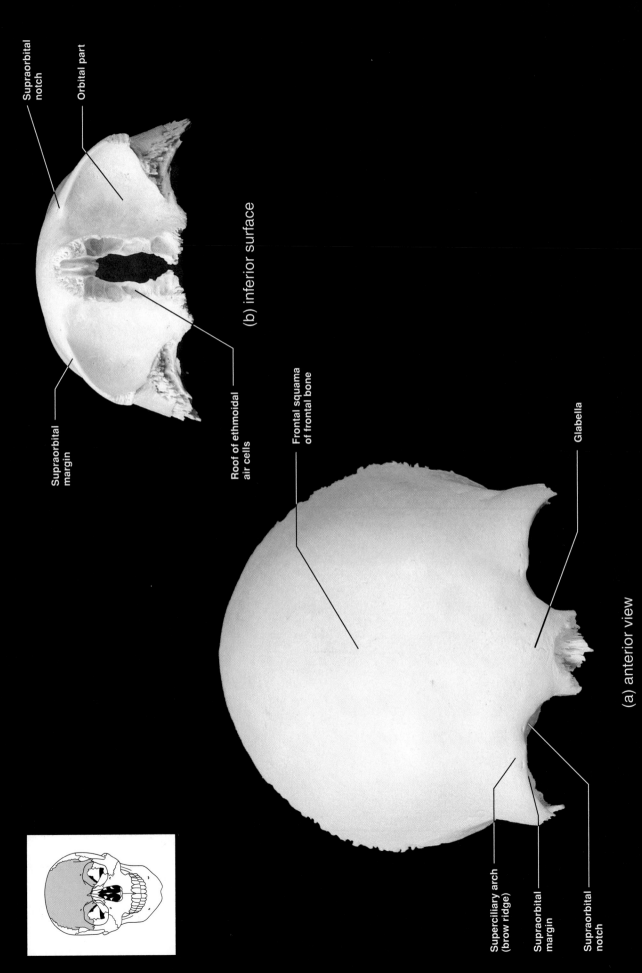

Supraorbital
notch

Orbital part

Supraorbital
margin

(b) inferior surface

Roof of ethmoidal
air cells

Frontal squama
of frontal bone

Glabella

Superciliary arch
(brow ridge)

Supraorbital
margin

Supraorbital
notch

(a) anterior view

Frontal bone.

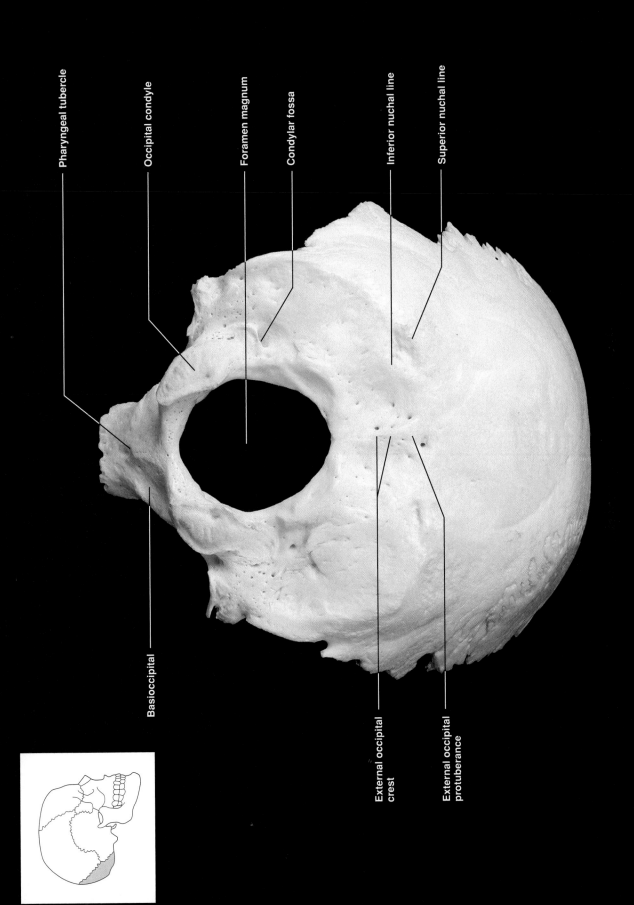

Pharyngeal tubercle

Occipital condyle

Foramen magnum

Condylar fossa

Inferior nuchal line

Superior nuchal line

Basioccipital

External occipital crest

External occipital protuberance

Occipital bone, inferior external view.

From *A Brief Atlas of the Human Body*, Second Edition. Matt Hutchinson, Jon Mallatt, Elaine N. Marieb, and Patricia Brady Wilhelm. Copyright © 2007 by Pearson Education, Inc. Published by Pearson Benjamin Cummings. All rights reserved.

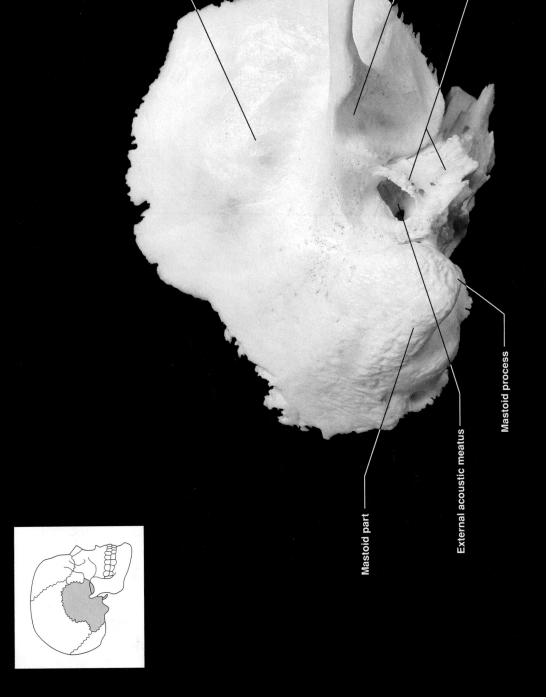

Squamous part

Zygomatic process

Mandibular fossa

Tympanic part

Mastoid process

Mastoid part

External acoustic meatus

(a) right lateral surface

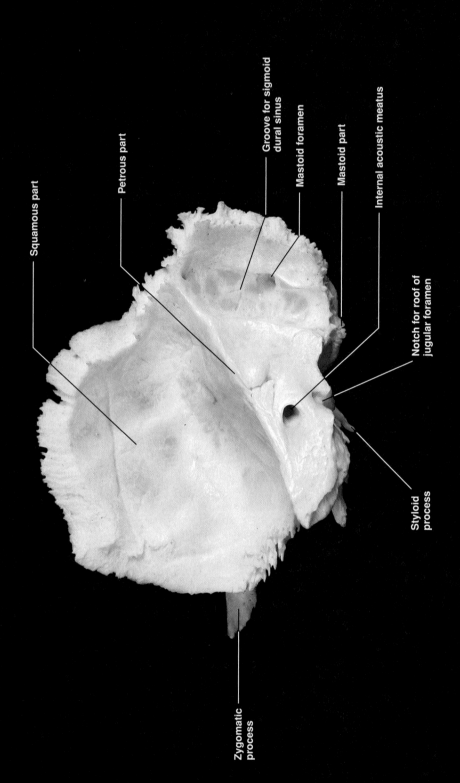

Squamous part

Petrous part

Groove for sigmoid dural sinus

Mastoid foramen

Mastoid part

Internal acoustic meatus

Notch for roof of jugular foramen

Styloid process

Zygomatic process

(b) right medial view

Temporal bone.

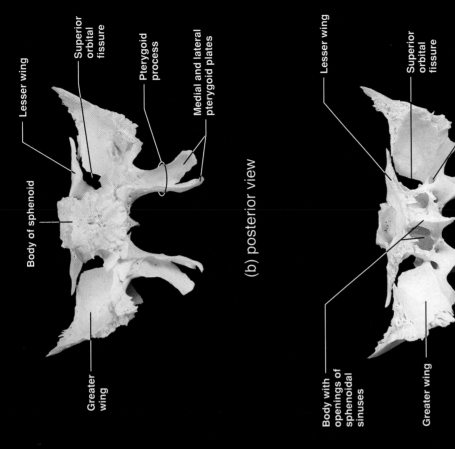

Lesser wing

Superior orbital fissure

Pterygoid process

Medial and lateral pterygoid plates

Body of sphenoid

Greater wing

(b) posterior view

Lesser wing

Superior orbital fissure

Foramen rotundum

Pterygoid process

Medial and lateral pterygoid plates

Body with openings of sphenoidal sinuses

Greater wing

(c) anterior view

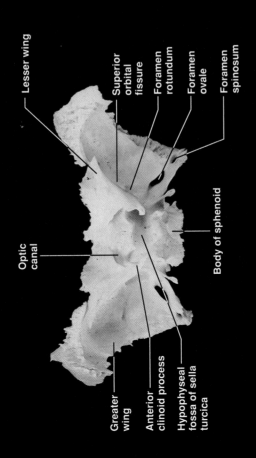

Lesser wing

Superior orbital fissure

Foramen rotundum

Foramen ovale

Foramen spinosum

Optic canal

Body of sphenoid

Greater wing

Anterior clinoid process

Hypophyseal fossa of sella turcica

(a) superior view

Sphenoid bone.

From *A Brief Atlas of the Human Body*, Second Edition. Matt Hutchinson, Jon Mallatt, Elaine N. Marieb, and Patricia Brady Wilhelm. Copyright © 2007 by Pearson Education, Inc. Published by Pearson Benjamin Cummings. All rights reserved.

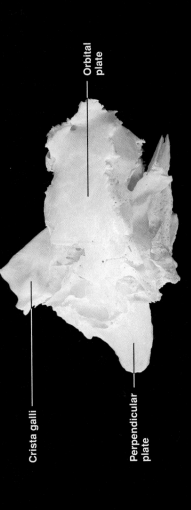

Crista galli

Perpendicular plate

Orbital plate

(a) left lateral surface

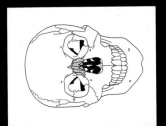

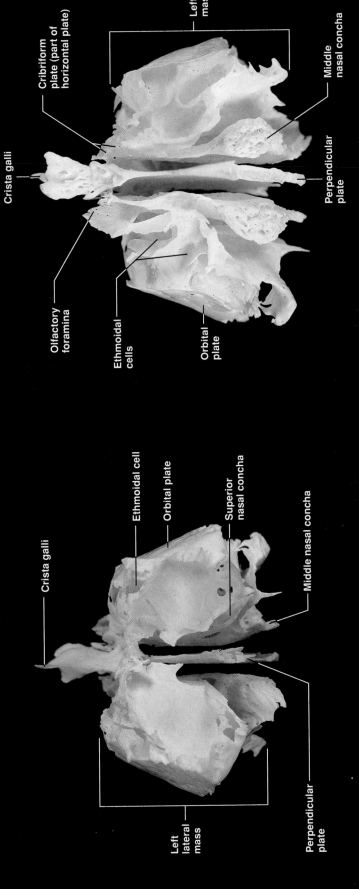

Crista galli

Olfactory foramina

Ethmoidal cells

Orbital plate

Cribriform plate (part of horizontal plate)

Left lateral mass

Middle nasal concha

Perpendicular plate

(c) anterior view

Crista galli

Ethmoidal cell

Orbital plate

Superior nasal concha

Middle nasal concha

Left lateral mass

Perpendicular plate

(b) posterior view

Ethmoid bone.

From *A Brief Atlas of the Human Body*, Second Edition. Matt Hutchinson, Jon Mallatt, Elaine N. Marieb, and Patricia Brady Wilhelm. Copyright © 2007 by Pearson Education, Inc. Published by Pearson Benjamin Cummings. All rights reserved.

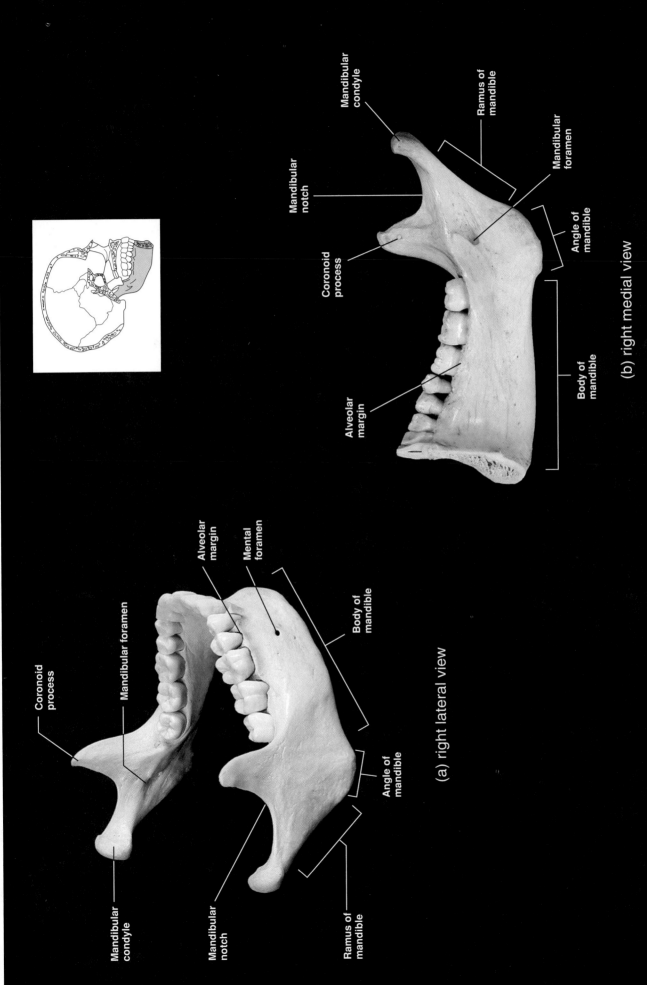

Coronoid process

Mandibular foramen

Mandibular condyle

Mandibular notch

Ramus of mandible

Alveolar margin

Mental foramen

Body of mandible

Angle of mandible

(a) right lateral view

Mandibular condyle

Mandibular notch

Coronoid process

Ramus of mandible

Mandibular foramen

Angle of mandible

Body of mandible

Alveolar margin

(b) right medial view

Mandible.

From *A Brief Atlas of the Human Body*, Second Edition. Matt Hutchinson, Jon Mallatt, Elaine N. Marieb, and Patricia Brady Wilhelm. Copyright © 2007 by Pearson Education, Inc. Published by Pearson Benjamin Cummings.

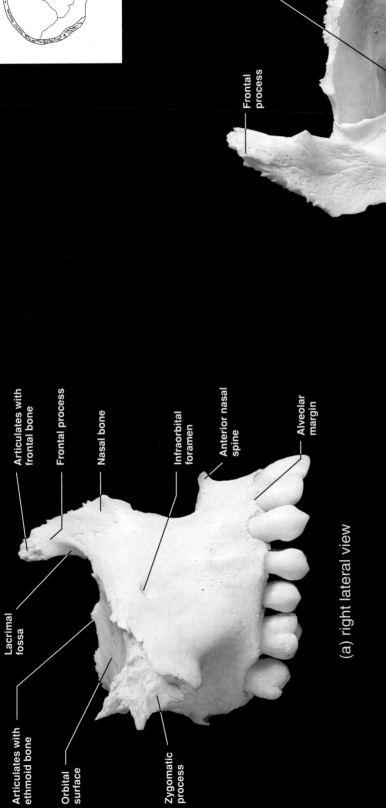

Articulates with ethmoid bone

Lacrimal fossa

Orbital surface

Zygomatic process

Articulates with frontal bone

Frontal process

Nasal bone

Infraorbital foramen

Anterior nasal spine

Alveolar margin

(a) right lateral view

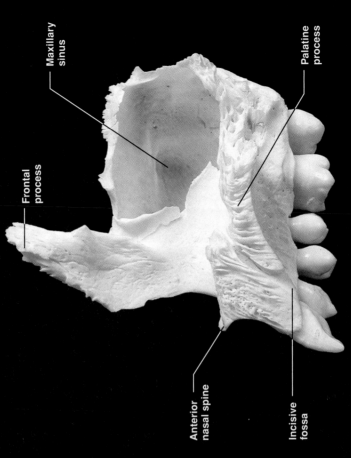

Frontal process

Maxillary sinus

Palatine process

Anterior nasal spine

Incisive fossa

(b) right medial view

Maxilla.

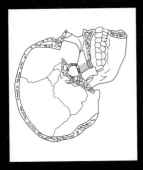

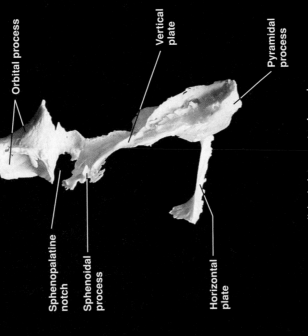

Orbital process

Vertical plate

Pyramidal process

Sphenopalatine notch

Sphenoidal process

Horizontal plate

(b) right posterior view

Palatine bone.

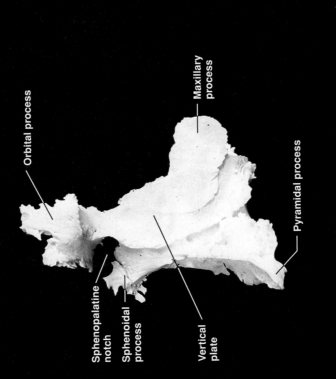

Orbital process

Maxillary process

Sphenopalatine notch

Sphenoidal process

Pyramidal process

Vertical plate

(a) right lateral view

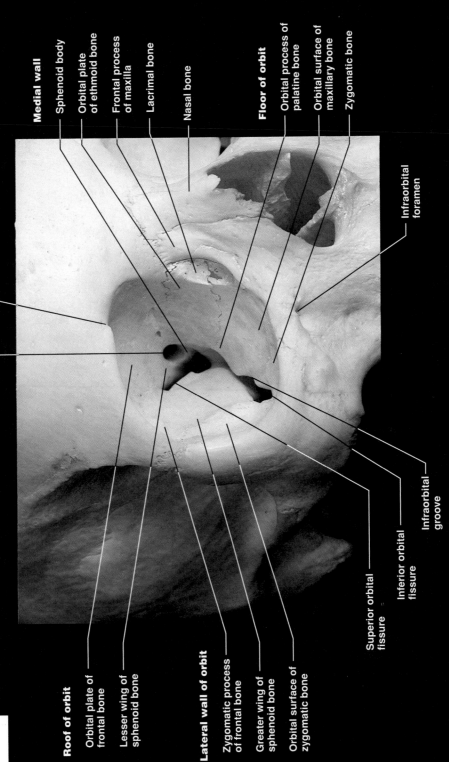

Medial wall

Sphenoid body

Orbital plate
of ethmoid bone

Frontal process
of maxilla

Lacrimal bone

Nasal bone

Floor of orbit

Orbital process of
palatine bone

Orbital surface of
maxillary bone

Zygomatic bone

Infraorbital
foramen

Supraorbital notch

Optic canal

Roof of orbit

Orbital plate of
frontal bone

Lesser wing of
sphenoid bone

Lateral wall of orbit

Zygomatic process
of frontal bone

Greater wing of
sphenoid bone

Orbital surface of
zygomatic bone

Superior orbital
fissure

Inferior orbital
fissure

Infraorbital
groove

Bony orbit.

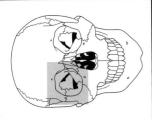

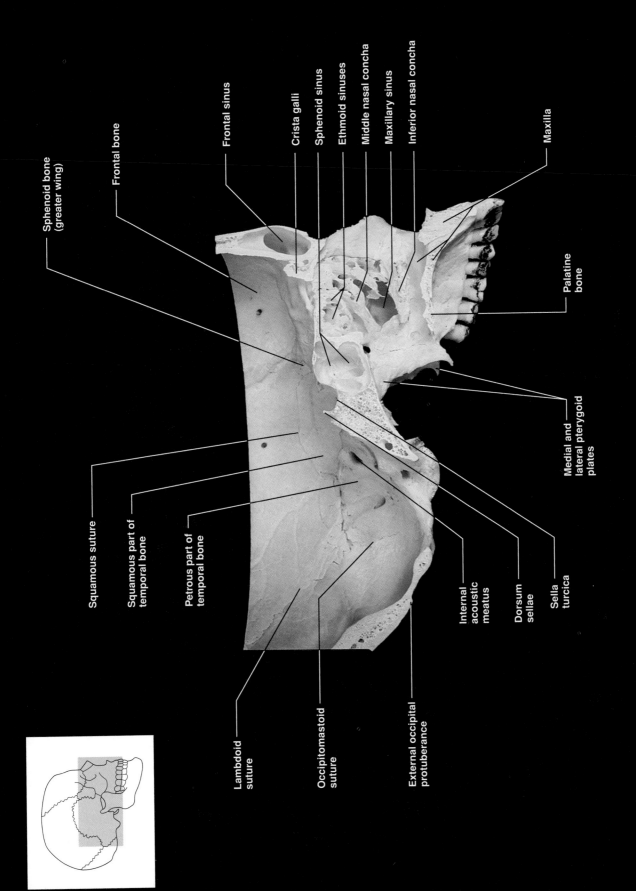

Sphenoid bone (greater wing)

Frontal bone

Frontal sinus

Crista galli

Sphenoid sinus

Ethmoid sinuses

Middle nasal concha

Maxillary sinus

Inferior nasal concha

Maxilla

Palatine bone

Medial and lateral pterygoid plates

Squamous suture

Squamous part of temporal bone

Petrous part of temporal bone

Internal acoustic meatus

Dorsum sellae

Sella turcica

Lambdoid suture

Occipitomastoid suture

External occipital protuberance

Nasal cavity, left lateral wall.

From *A Brief Atlas of the Human Body*, Second Edition. Matt Hutchinson, Jon Mallatt, Elaine N. Marieb, and Patricia Brady Wilhelm. Copyright © 2007 by Pearson Education, Inc. Published by Pearson Benjamin Cummings. All rights reserved.

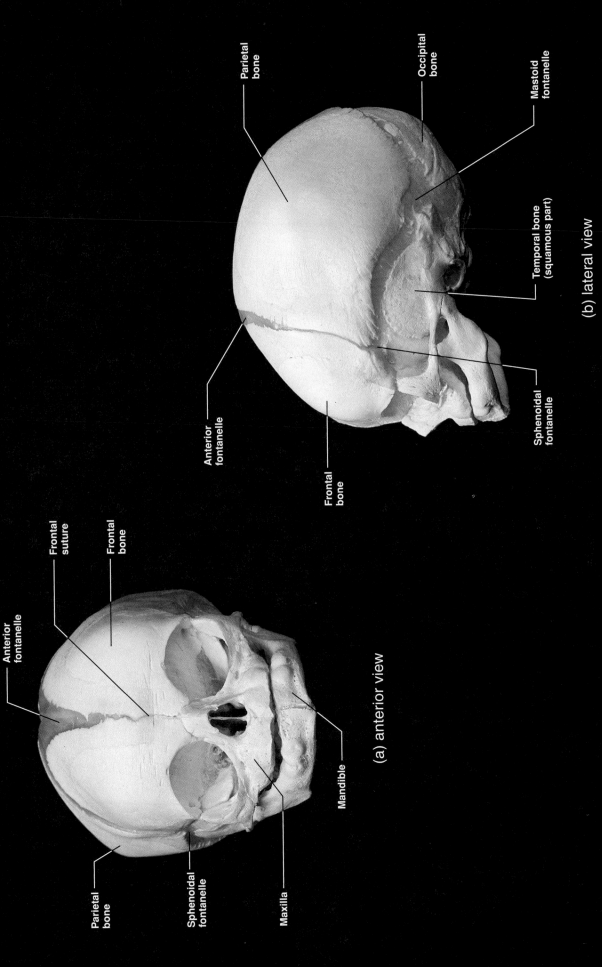

Parietal
bone

Anterior
fontanelle

Frontal
suture

Frontal
bone

Sphenoidal
fontanelle

Maxilla

Mandible

(a) anterior view

Anterior
fontanelle

Parietal
bone

Occipital
bone

Mastoid
fontanelle

Frontal
bone

Temporal bone
(squamous part)

Sphenoidal
fontanelle

(b) lateral view

Fetal skull.

From *A Brief Atlas of the Human Body*, Second Edition. Matt Hutchinson, Jon Mallatt, Elaine N. Marieb, and Patricia Brady Wilhelm. Copyright © 2007 by Pearson Education, Inc. Published by Pearson Benjamin Cummings. All rights reserved.

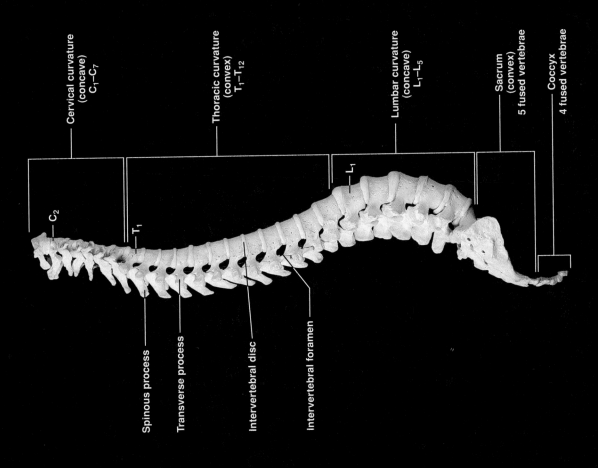

Cervical curvature (concave) C_1–C_7

Thoracic curvature (convex) T_1–T_{12}

Lumbar curvature (concave) L_1–L_5

Sacrum (convex) 5 fused vertebrae

Coccyx 4 fused vertebrae

C_2

T_1

L_1

Spinous process

Transverse process

Intervertebral disc

Intervertebral foramen

(a) right lateral view

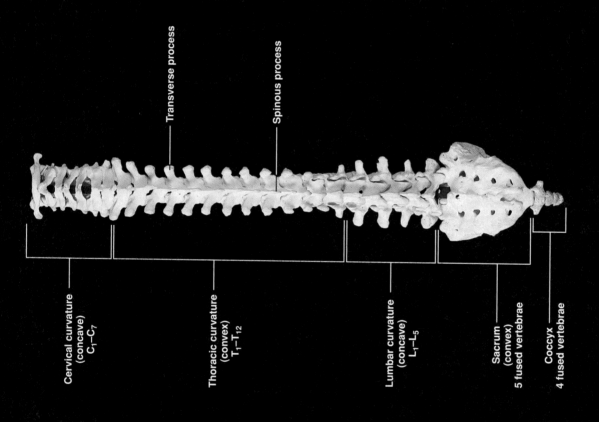

Transverse process

Spinous process

Cervical curvature
(concave)
$C_1–C_7$

Thoracic curvature
(convex)
$T_1–T_{12}$

Lumbar curvature
(concave)
$L_1–L_5$

Sacrum
(convex)
5 fused vertebrae

Coccyx
4 fused vertebrae

(b) posterior view

Articulated vertebral column.

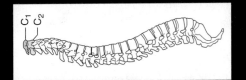

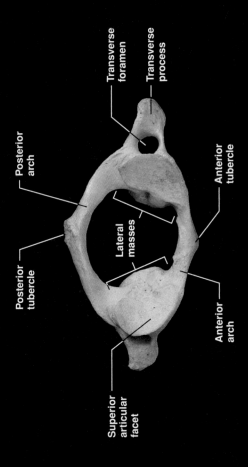

Posterior
arch

Transverse
foramen

Transverse
process

Posterior
tubercle

Lateral
masses

Anterior
tubercle

Superior
articular
facet

Anterior
arch

(a) atlas, superior view

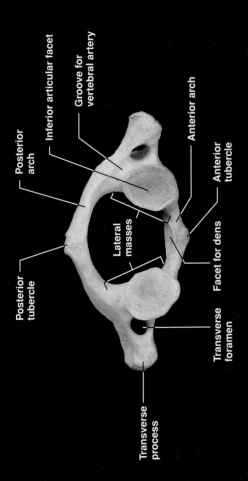

Posterior
arch

Inferior articular facet

Groove for
vertebral artery

Anterior arch

Posterior
tubercle

Lateral
masses

Anterior
tubercle

Transverse
foramen

Facet for dens

Transverse
process

(b) atlas, inferior view

From *A Brief Atlas of the Human Body*, Second Edition. Matt Hutchinson, Jon Mallatt, Elaine N. Marieb, and Patricia Brady Wilhelm. Copyright © 2007 by Pearson Education, Inc. Published by Pearson Benjamin Cummings. All rights reserved.

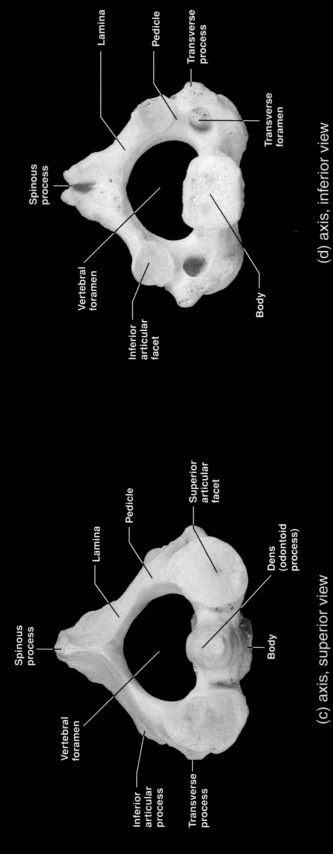

Spinous process

Lamina

Pedicle

Transverse process

Transverse foramen

Vertebral foramen

Inferior articular facet

Body

(d) axis, inferior view

Spinous process

Lamina

Pedicle

Superior articular facet

Dens (odontoid process)

Vertebral foramen

Body

Inferior articular process

Transverse process

(c) axis, superior view

Body of axis

Anterior arch of atlas

Dens (odontoid process)

(e) articulated atlas and axis, superior view

Various views of vertebrae C_1 and C_2.

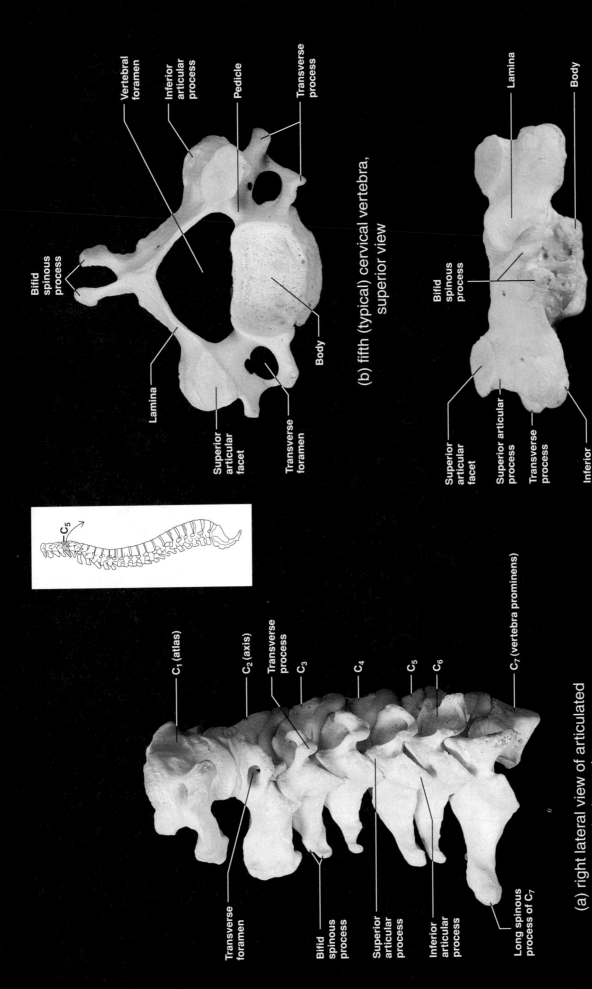

Vertebral foramen

Inferior articular process

Pedicle

Transverse process

Bifid spinous process

Lamina

Superior articular facet

Transverse foramen

Body

(b) fifth (typical) cervical vertebra, superior view

Lamina

Body

Bifid spinous process

Superior articular facet

Superior articular process

Transverse process

Inferior articular process

(c) fifth (typical) cervical vertebra, posterior view

C_5

C_1 (atlas)

C_2 (axis)

Transverse process

C_3

C_4

C_5

C_6

C_7 (vertebra prominens)

Transverse foramen

Bifid spinous process

Superior articular process

Inferior articular process

Long spinous process of C_7

(a) right lateral view of articulated cervical vertebrae

From *A Brief Atlas of the Human Body*, Second Edition. Matt Hutchinson, Jon Mallatt, Elaine N. Marieb, and Patricia Brady Wilhelm. Copyright © 2007 by Pearson Education, Inc. Published by Pearson Benjamin Cummings. All rights reserved.

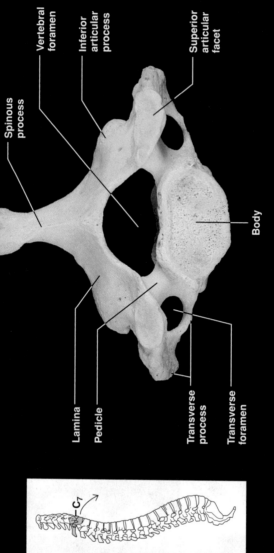

Spinous process

Vertebral foramen

Inferior articular process

Superior articular facet

Body

Lamina

Pedicle

Transverse process

Transverse foramen

(e) vertebra prominens (C₇), superior view

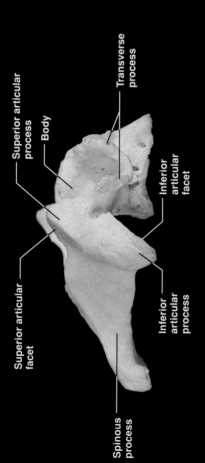

Superior articular process

Body

Transverse process

Inferior articular facet

Superior articular facet

Inferior articular process

Spinous process

(d) fifth (typical) cervical vertebra, right lateral view

C₇

Cervical vertebrae.

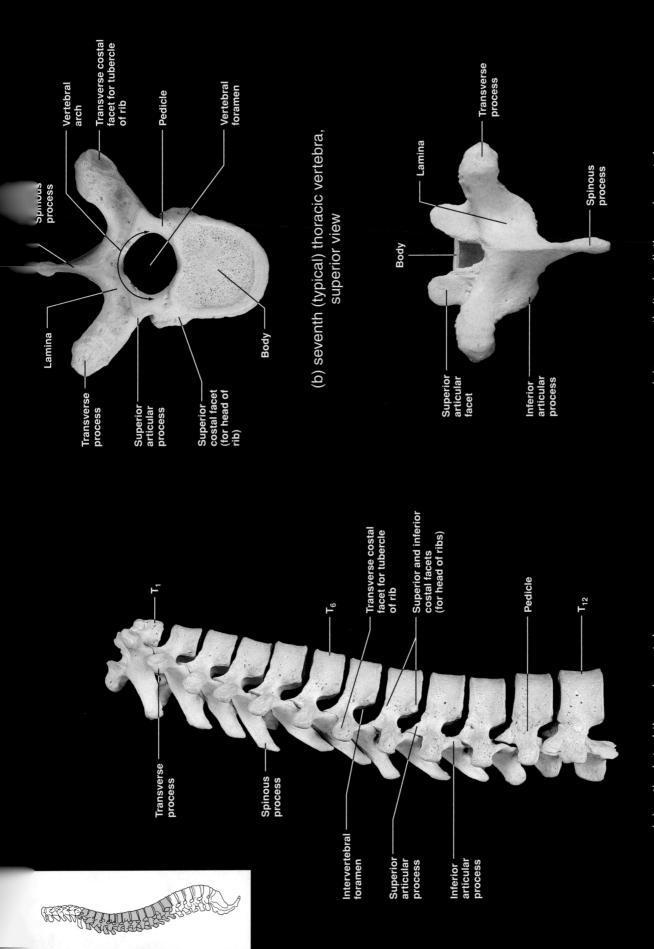

Spinous process

Vertebral arch

Transverse costal facet for tubercle of rib

Pedicle

Vertebral foramen

Lamina

Transverse process

Superior articular process

Superior costal facet (for head of rib)

Body

(b) seventh (typical) thoracic vertebra, superior view

Transverse process

Lamina

Body

Spinous process

Superior articular facet

Inferior articular process

(c) seventh (typical) thoracic vertebra, posterior view

T₁

T₆

T₁₂

Transverse costal facet for tubercle of rib

Superior and inferior costal facets (for head of ribs)

Pedicle

Transverse process

Spinous process

Intervertebral foramen

Superior articular process

Inferior articular process

(a) articulated thoracic vertebrae, right lateral view

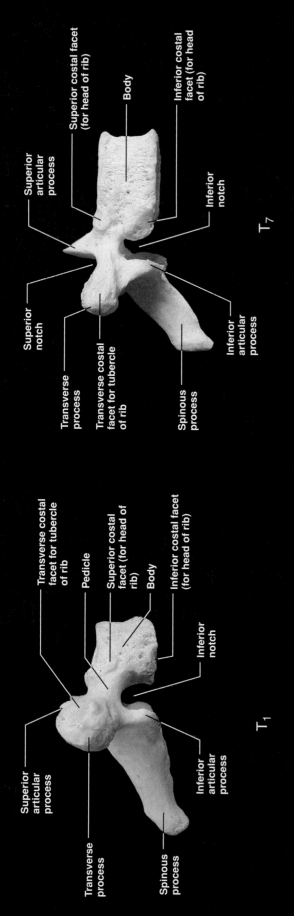

T₁

- Superior articular process
- Transverse costal facet for tubercle of rib
- Pedicle
- Superior costal facet (for head of rib)
- Body
- Inferior costal facet (for head of rib)
- Inferior notch
- Transverse process
- Spinous process
- Inferior articular process

Superior articular process

Superior costal facet (for head of rib)

Body

Inferior costal facet (for head of rib)

Inferior notch

Superior notch

Transverse process

Transverse costal facet for tubercle of rib

Spinous process

Inferior articular process

T₇

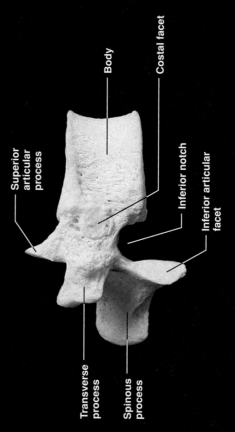

- Superior articular process
- Body
- Costal facet
- Inferior notch
- Inferior articular facet
- Transverse process
- Spinous process

T₁₂

(d) comparison of T₁, T₇, and T₁₂ in right lateral views

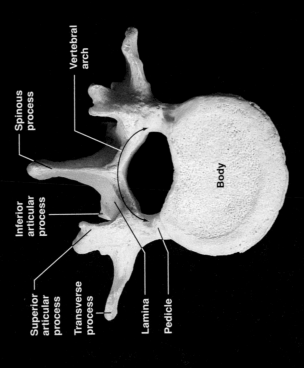

Spinous process

Vertebral arch

Inferior articular process

Superior articular process

Transverse process

Lamina

Pedicle

Body

(b) second lumbar vertebra, superior view

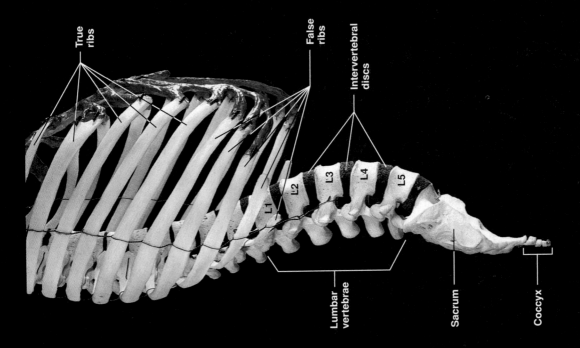

True ribs

False ribs

Intervertebral discs

L1

L2

L3

L4

L5

Lumbar vertebrae

Sacrum

Coccyx

(a) articulated lumbar vertebrae and rib cage, right lateral view

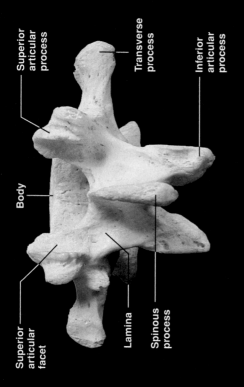

Superior articular process

Transverse process

Inferior articular process

Body

Superior articular facet

Lamina

Spinous process

(c) second lumbar vertebra, posterior view

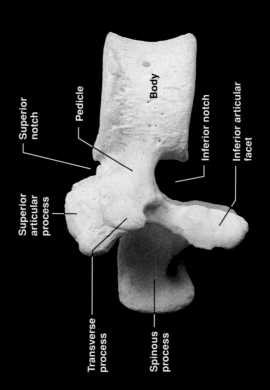

Superior notch

Pedicle

Body

Inferior notch

Inferior articular facet

Superior articular process

Transverse process

Spinous process

(d) second lumbar vertebra, right lateral view

Lumbar vertebrae.

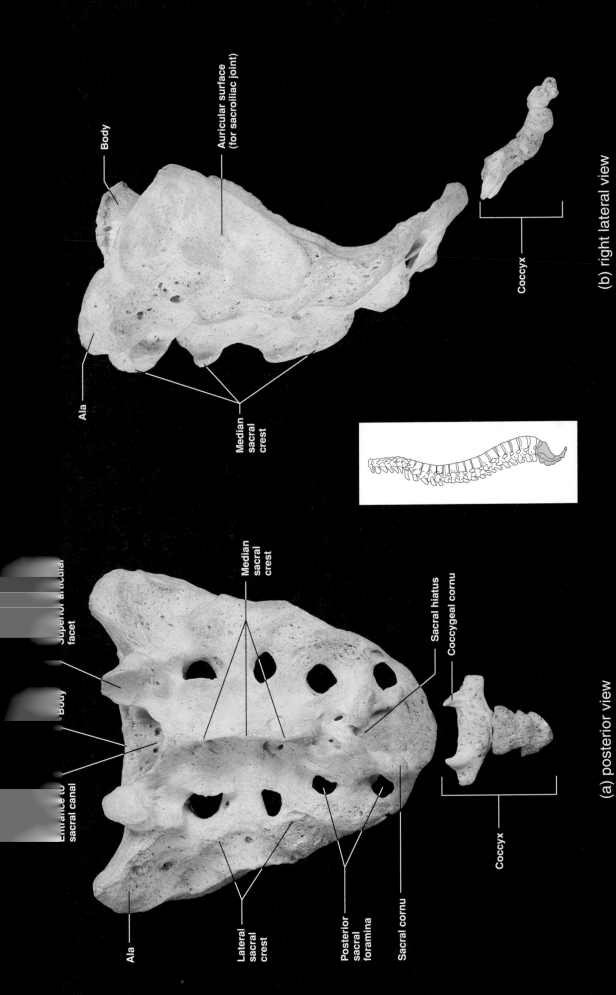

Body

Auricular surface
(for sacroiliac joint)

Ala

Median
sacral
crest

Coccyx

(b) right lateral view

Median
sacral
crest

Superior articular
facet

Body

Entrance to
sacral canal

Sacral hiatus
Coccygeal cornu

Ala

Lateral
sacral
crest

Posterior
sacral
foramina

Sacral cornu

Coccyx

(a) posterior view

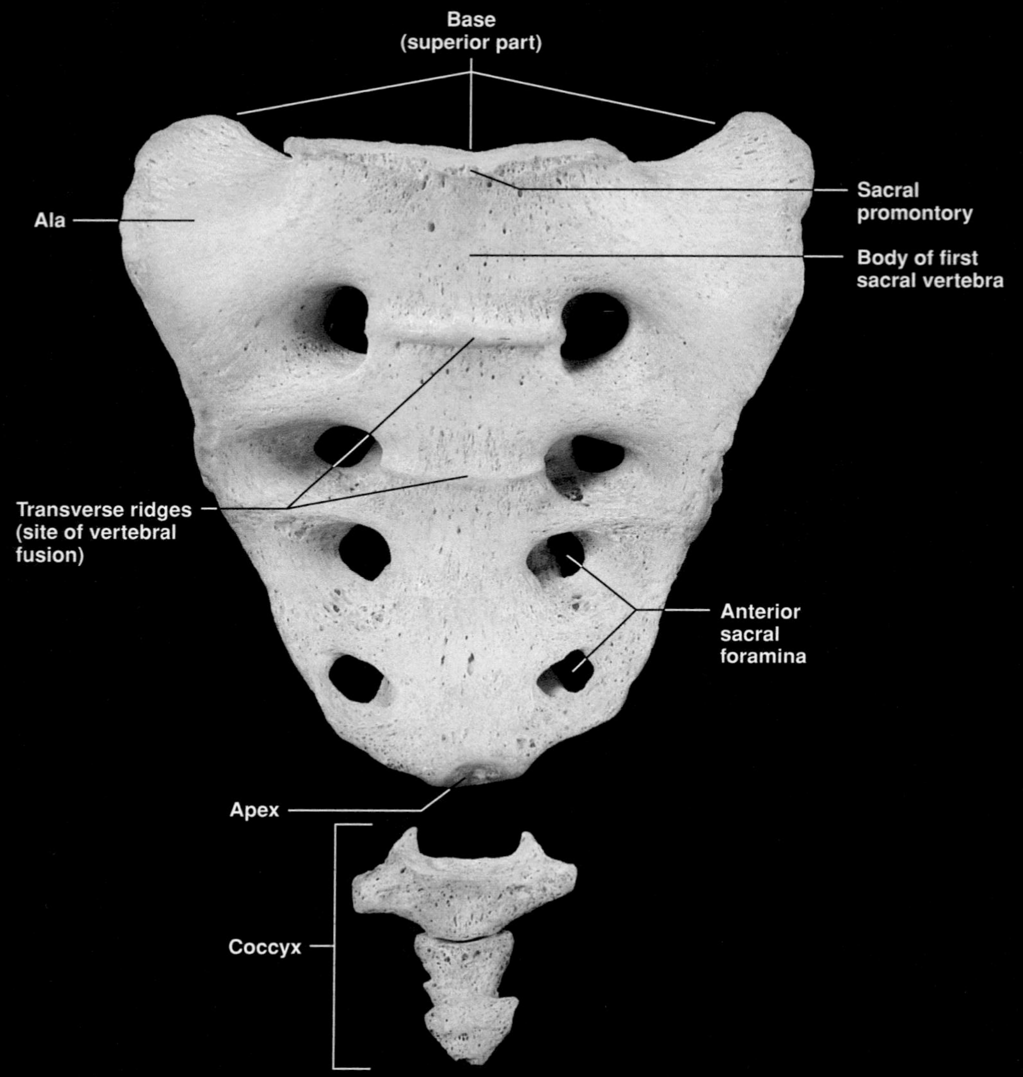

Sacral
promontory

Body of first
sacral vertebra

Base
(superior part)

Anterior
sacral
foramina

Ala

Transverse ridges
(site of vertebral
fusion)

Apex

Coccyx

(c) anterior view

Sacrum and coccyx.

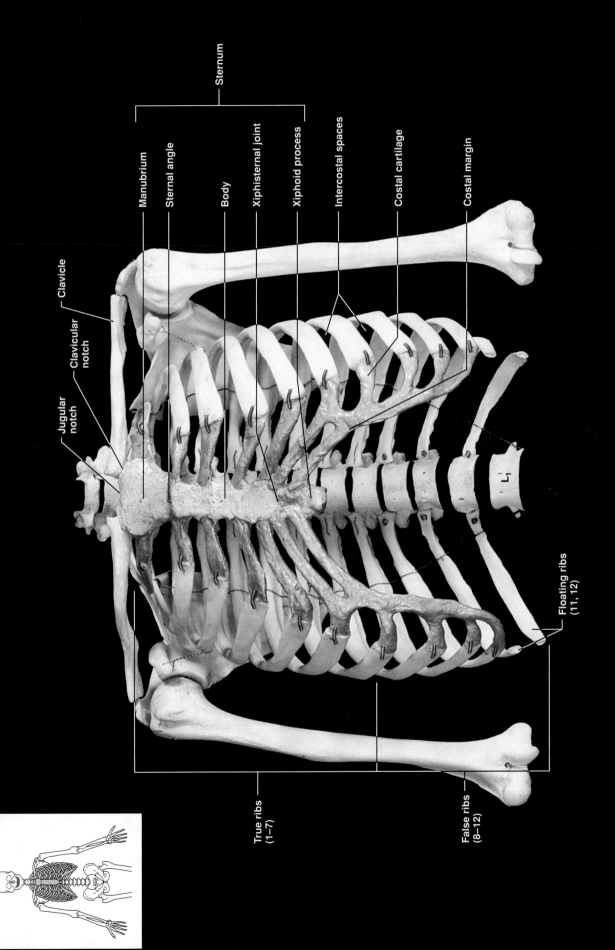

Clavicle

Clavicular notch

Jugular notch

Sternum

Manubrium

Sternal angle

Body

Xiphisternal joint

Xiphoid process

Intercostal spaces

Costal cartilage

Costal margin

L₁

True ribs (1–7)

False ribs (8–12)

Floating ribs (11, 12)

(a) anterior view

From *A Brief Atlas of the Human Body*, Second Edition. Matt Hutchinson, Jon Mallatt, Elaine N. Marieb, and Patricia Brady Wilhelm. Copyright © 2007 by Pearson Education, Inc. Published by Pearson Benjamin Cummings. All rights reserved.

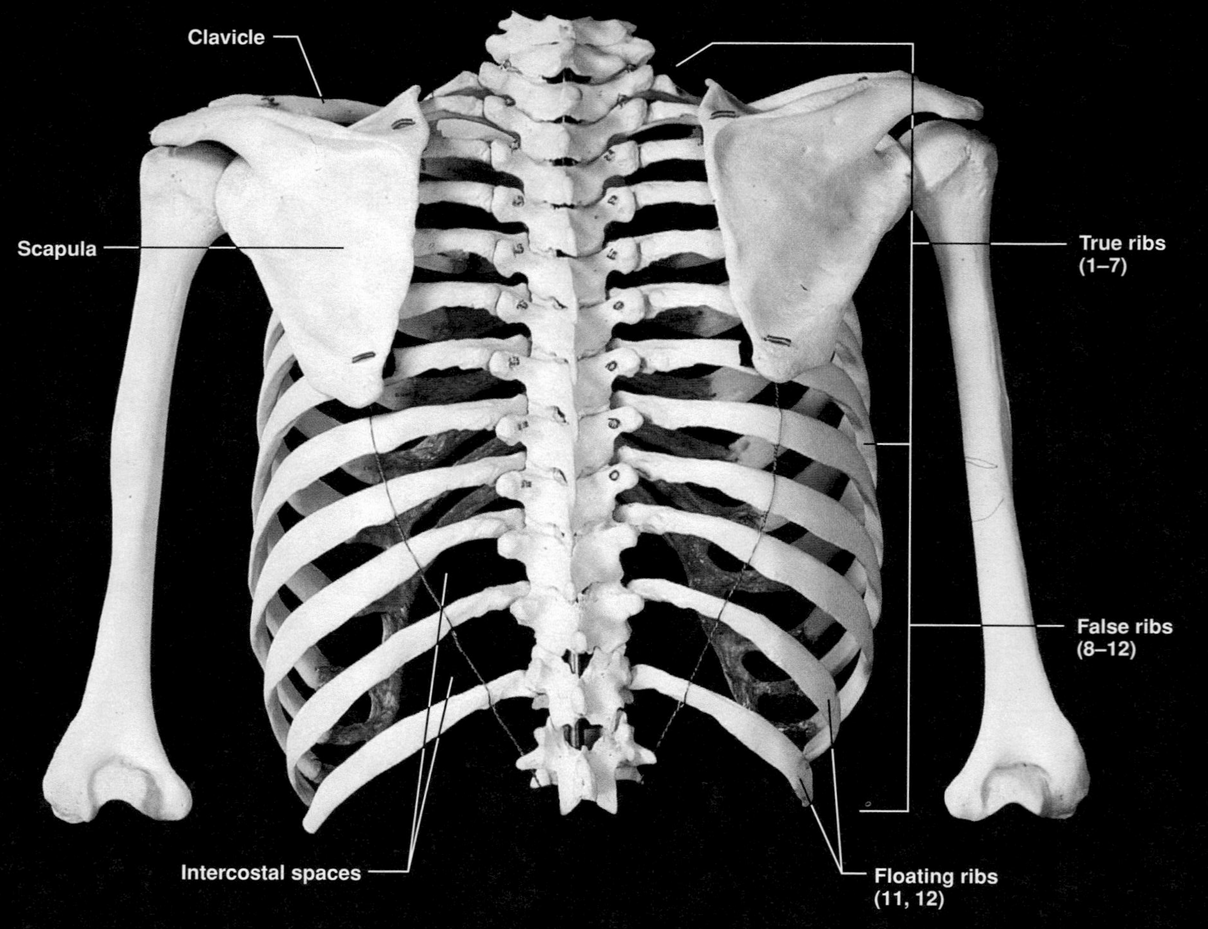

True ribs
(1–7)

False ribs
(8–12)

Floating ribs
(11, 12)

Clavicle

Scapula

Intercostal spaces

(b) posterior view

Thoracic cage.

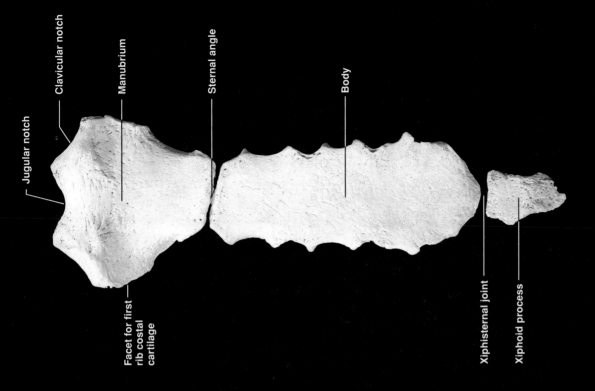

Jugular notch

Clavicular notch

Manubrium

Sternal angle

Body

Facet for first rib costal cartilage

Xiphisternal joint

Xiphoid process

(d) sternum, anterior view

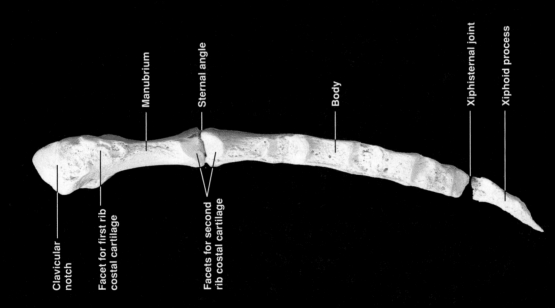

Clavicular notch

Facet for first rib costal cartilage

Manubrium

Sternal angle

Facets for second rib costal cartilage

Body

Xiphisternal joint

Xiphoid process

(c) sternum, right lateral view

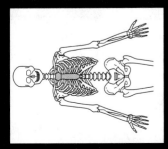

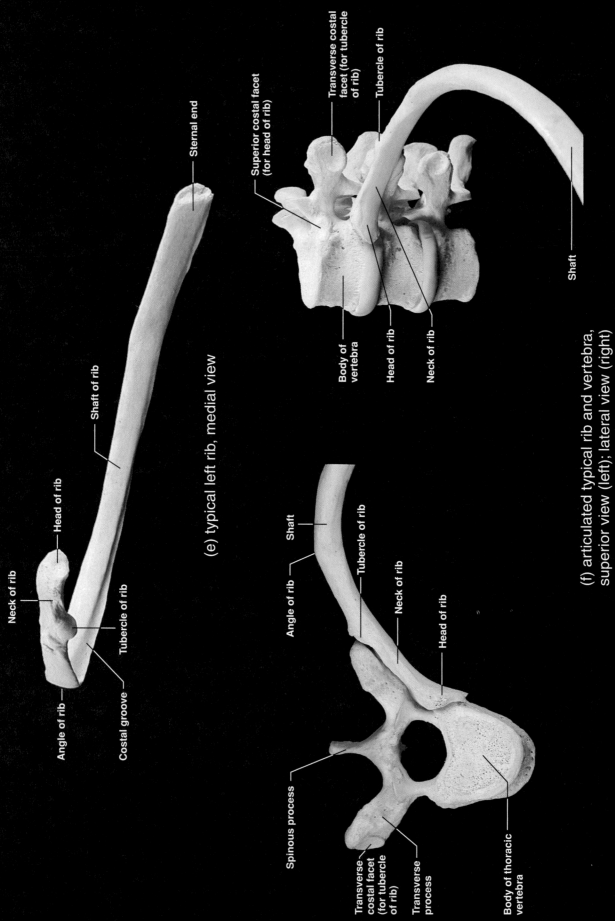

Neck of rib

Head of rib

Shaft of rib

Angle of rib

Costal groove

Tubercle of rib

Sternal end

(e) typical left rib, medial view

Superior costal facet
(for head of rib)

Transverse costal
facet (for tubercle
of rib)

Tubercle of rib

Body of
vertebra

Head of rib

Neck of rib

Shaft

Angle of rib

Shaft

Tubercle of rib

Neck of rib

Head of rib

Spinous process

Transverse
costal facet
(for tubercle
of rib)

Transverse
process

Body of thoracic
vertebra

(f) articulated typical rib and vertebra,
superior view (left); lateral view (right)

Thoracic cage (continued).

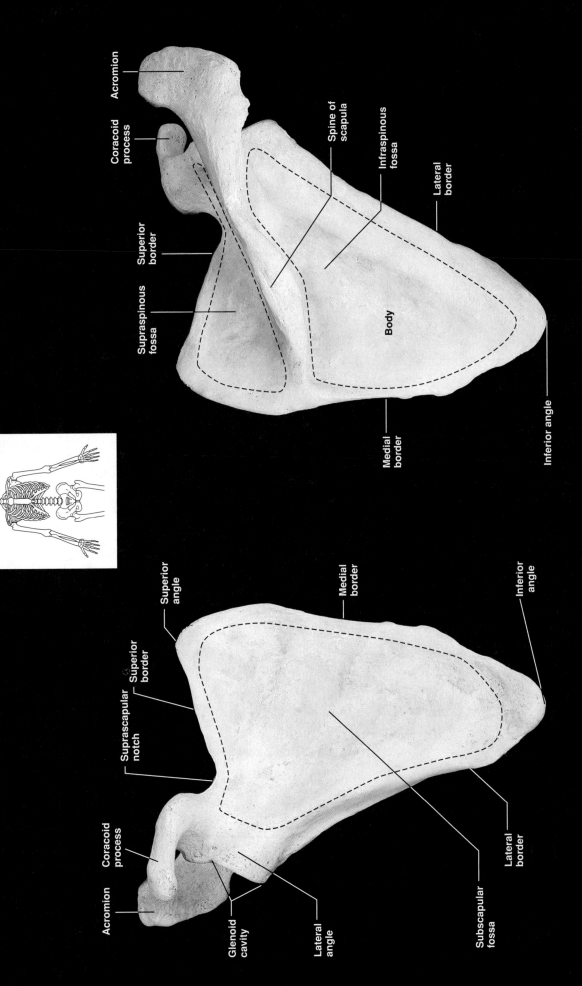

Acromion

Coracoid process

Superior border

Supraspinous fossa

Spine of scapula

Infraspinous fossa

Lateral border

Body

Medial border

Inferior angle

(b) right scapula, posterior view

Superior angle

Superior border

Suprascapular notch

Medial border

Inferior angle

Acromion

Coracoid process

Glenoid cavity

Lateral angle

Subscapular fossa

Lateral border

(a) right scapula, anterior view

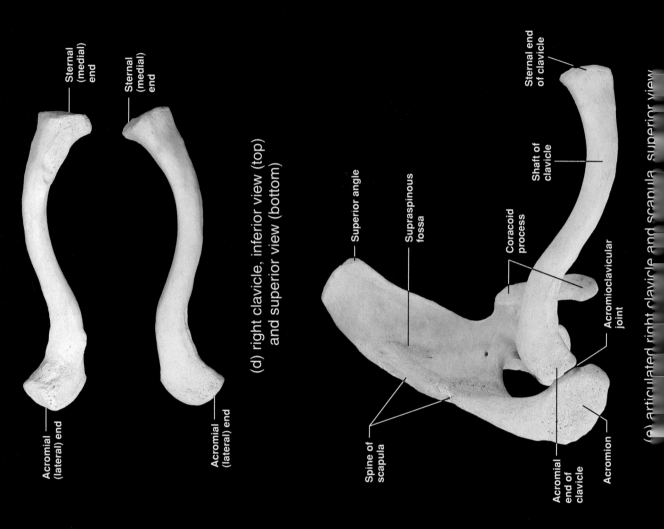

Sternal (medial) end

Sternal (medial) end

Acromial (lateral) end

Acromial (lateral) end

(d) right clavicle, inferior view (top) and superior view (bottom)

Superior angle

Supraspinous fossa

Coracoid process

Sternal end of clavicle

Shaft of clavicle

Acromioclavicular joint

Spine of scapula

Acromial end of clavicle

Acromion

(e) articulated right clavicle and scapula, superior view

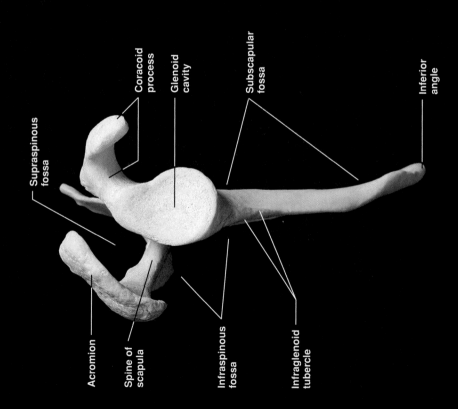

Supraspinous fossa

Coracoid process

Glenoid cavity

Subscapular fossa

Inferior angle

Acromion

Spine of scapula

Infraspinous fossa

Infraglenoid tubercle

(c) right scapula, lateral aspect

Scapula and clavicle

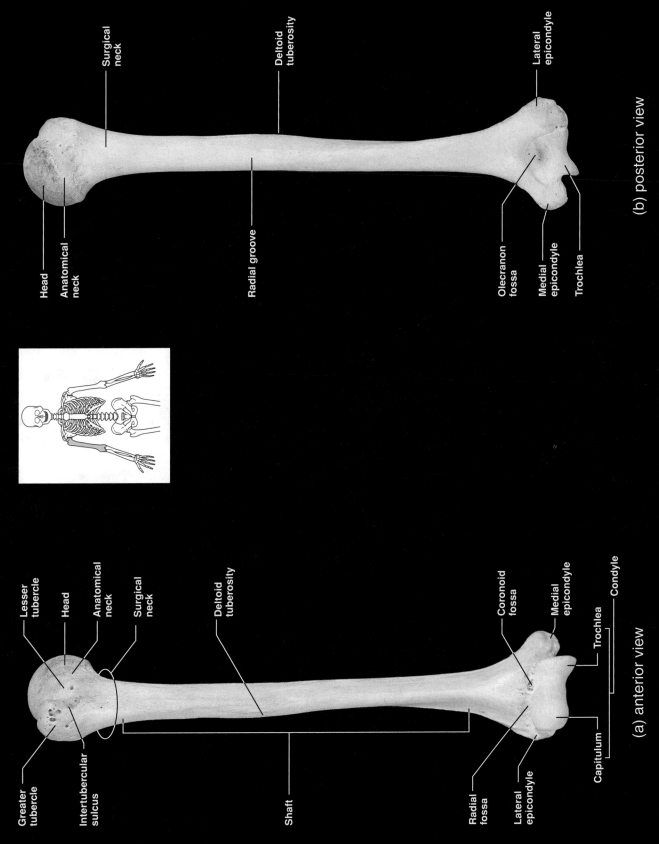

Lesser
tubercle

Head

Anatomical
neck

Surgical
neck

Deltoid
tuberosity

Greater
tubercle

Intertubercular
sulcus

Shaft

Coronoid
fossa

Medial
epicondyle

Trochlea

Condyle

Capitulum

Radial
fossa

Lateral
epicondyle

(a) anterior view

Surgical
neck

Deltoid
tuberosity

Lateral
epicondyle

Head

Anatomical
neck

Radial groove

Olecranon
fossa

Medial
epicondyle

Trochlea

(b) posterior view

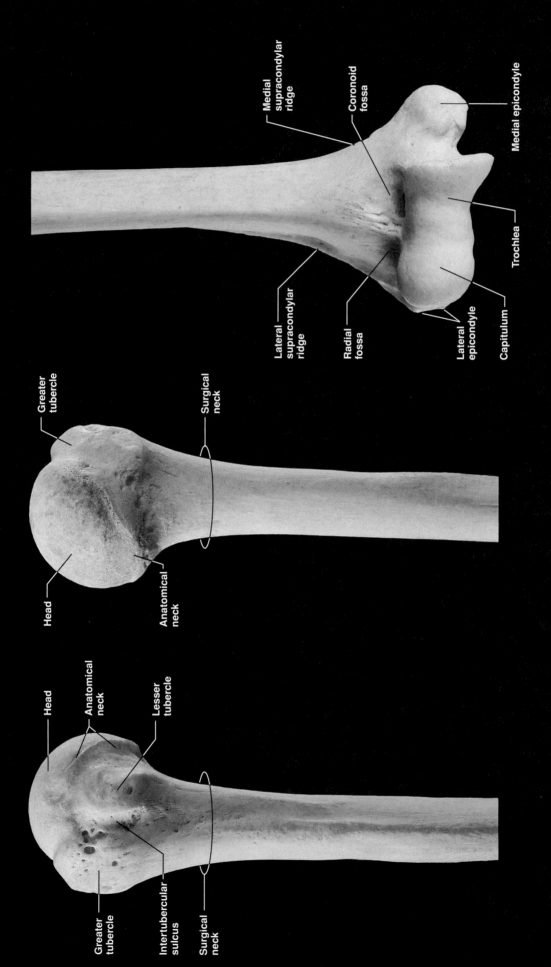

Head

Anatomical
neck

Lesser
tubercle

Greater
tubercle

Intertubercular
sulcus

Surgical
neck

(c) proximal end, anterior view

Greater
tubercle

Head

Anatomical
neck

Surgical
neck

(d) proximal end, posterior view

Medial supracondylar
ridge

Coronoid
fossa

Medial epicondyle

Lateral
supracondylar
ridge

Radial
fossa

Lateral
epicondyle

Capitulum

Trochlea

(e) distal end, anterior view

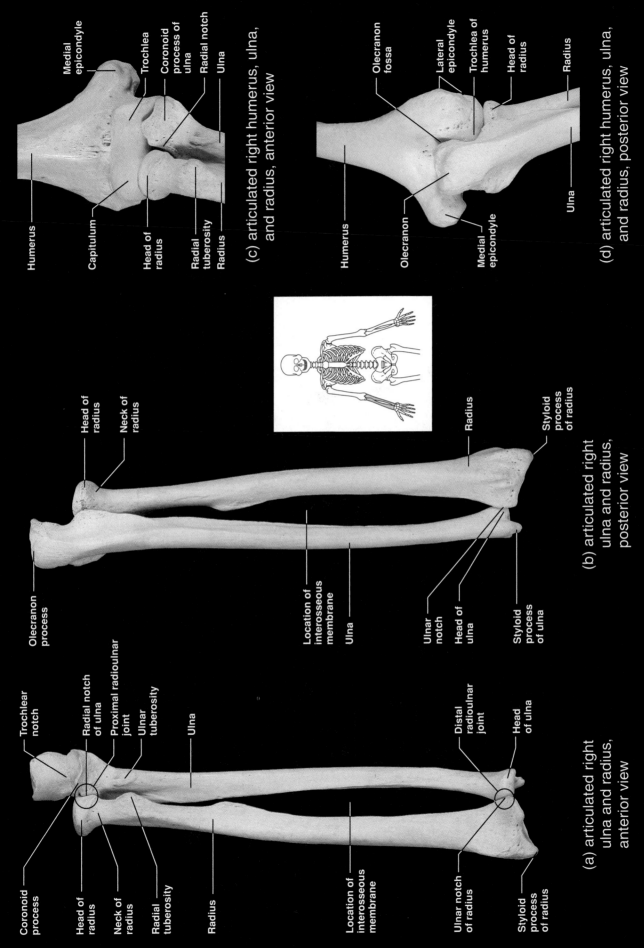

Medial
epicondyle

Trochlea

Coronoid
process of
ulna

Radial notch

Ulna

Humerus

Capitulum

Head of
radius

Radial
tuberosity

Radius

(c) articulated right humerus, ulna,
and radius, anterior view

Olecranon
fossa

Lateral
epicondyle

Trochlea of
humerus

Head of
radius

Radius

Humerus

Olecranon

Medial
epicondyle

Ulna

(d) articulated right humerus, ulna,
and radius, posterior view

Head of
radius

Neck of
radius

Olecranon
process

Location of
interosseous
membrane

Ulna

Ulnar
notch

Head of
ulna

Styloid
process
of ulna

Radius

Styloid
process
of radius

(b) articulated right
ulna and radius,
posterior view

Trochlear
notch

Radial notch
of ulna

Proximal radioulnar
joint

Ulnar
tuberosity

Ulna

Distal
radioulnar
joint

Head
of ulna

Coronoid
process

Head of
radius

Neck of
radius

Radial
tuberosity

Radius

Location of
interosseous
membrane

Ulnar notch
of radius

Styloid
process
of radius

(a) articulated right
ulna and radius,
anterior view

From *A Brief Atlas of the Human Body*, Second Edition. Matt Hutchinson, Jon Mallatt, Elaine N. Marieb, and Patricia Brady Wilhelm. Copyright © 2007 by Pearson Education, Inc. Published by Pearson Benjamin Cummings. All rights reserved.

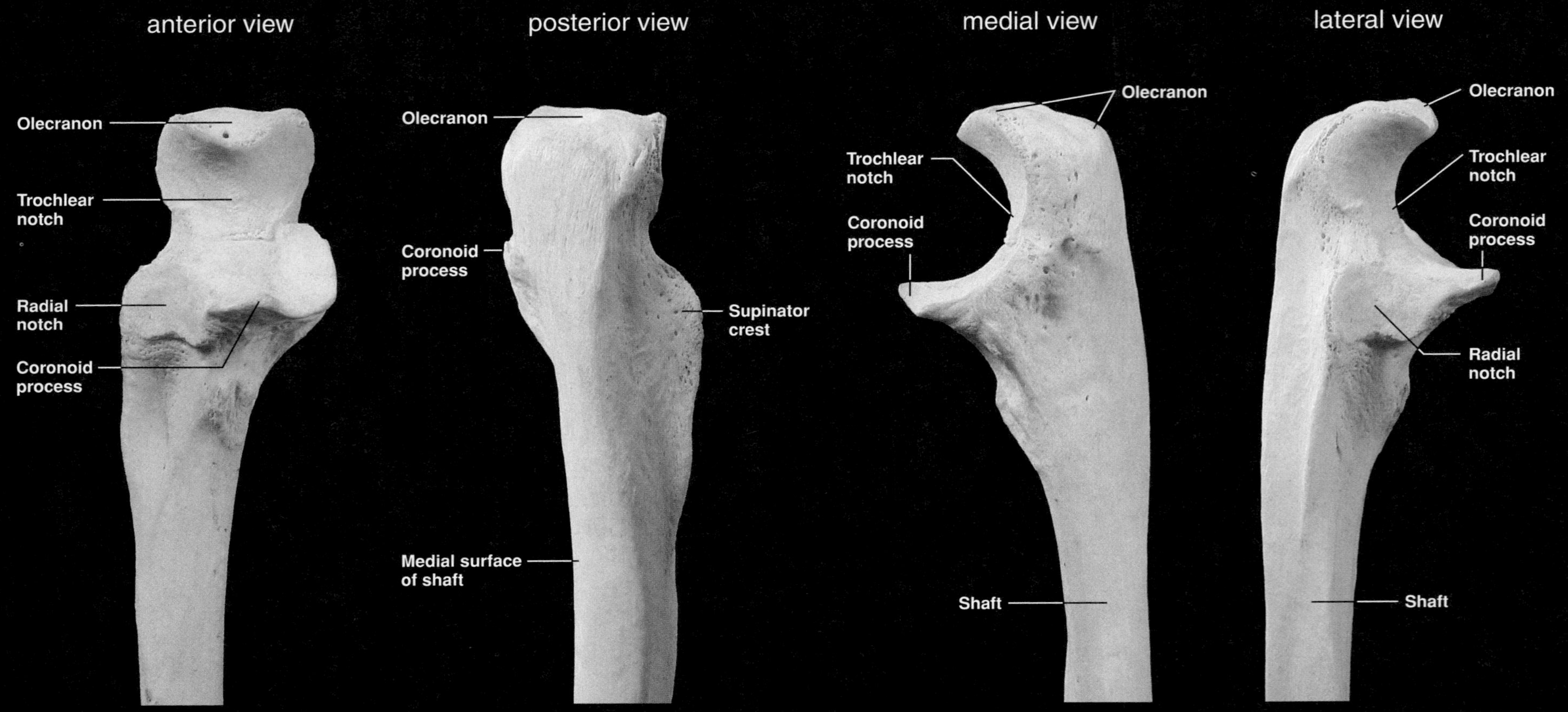

anterior view

Olecranon

Trochlear notch

Radial notch

Coronoid process

posterior view

Olecranon

Coronoid process

Supinator crest

Medial surface of shaft

medial view

Olecranon

Trochlear notch

Coronoid process

Shaft

lateral view

Olecranon

Trochlear notch

Coronoid process

Radial notch

Shaft

(e) right ulna, proximal end

Right ulna and radius.

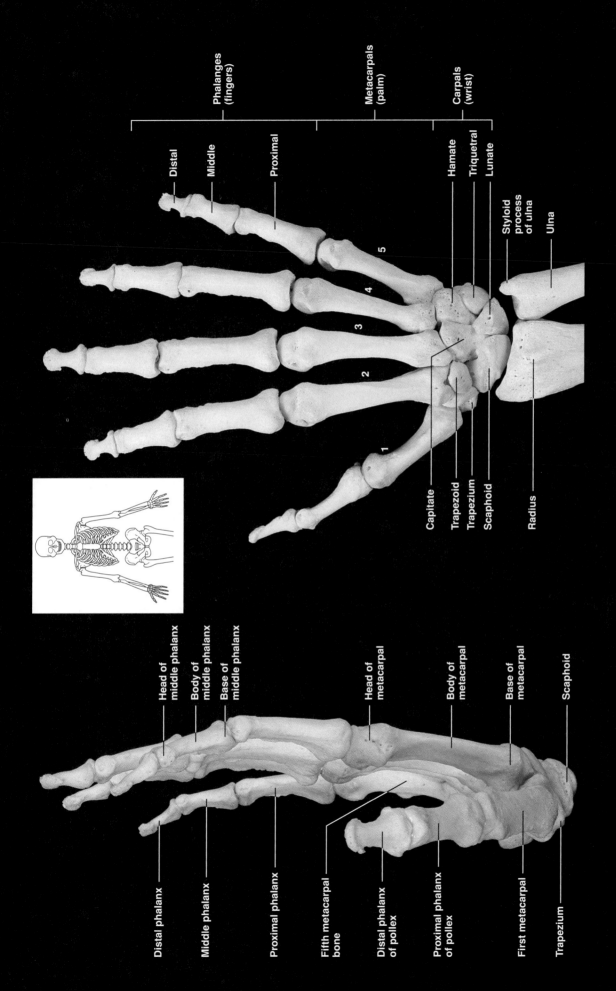

Phalanges (fingers)

Distal

Middle

Proximal

Metacarpals (palm)

Carpals (wrist)

Hamate

Triquetral

Lunate

Styloid process of ulna

Ulna

Capitate

Trapezoid

Trapezium

Scaphoid

Radius

(b) dorsal aspect

Head of middle phalanx

Body of middle phalanx

Base of middle phalanx

Head of metacarpal

Body of metacarpal

Base of metacarpal

Scaphoid

Distal phalanx

Middle phalanx

Proximal phalanx

Fifth metacarpal bone

Distal phalanx of pollex

Proximal phalanx of pollex

First metacarpal

Trapezium

(a) lateral aspect

Bones of the right hand.

From *A Brief Atlas of the Human Body*, Second Edition. Matt Hutchinson, Jon Mallatt, Elaine N. Marieb, and Patricia Brady Wilhelm. Copyright © 2007 by Pearson Education, Inc. Published by Pearson Benjamin Cummings. All rights reserved.

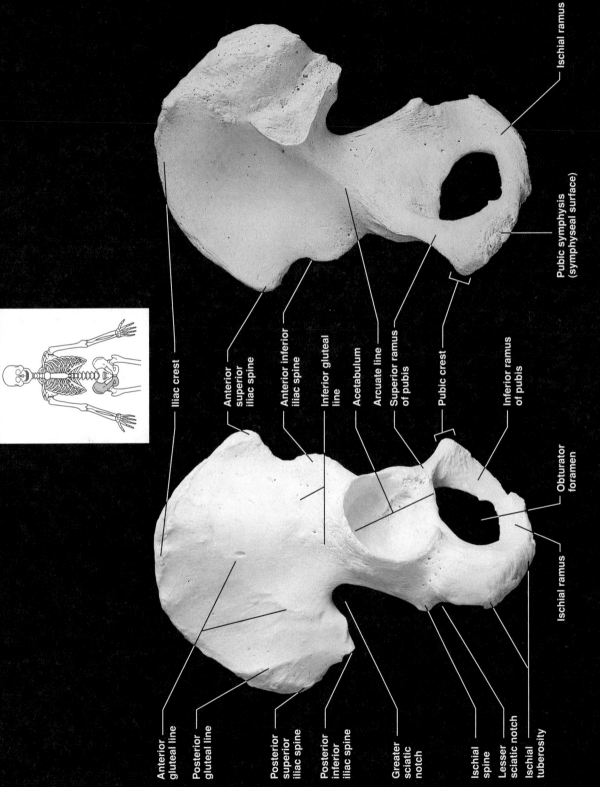

Anterior gluteal line

Posterior gluteal line

Posterior superior iliac spine

Posterior inferior iliac spine

Greater sciatic notch

Ischial spine

Lesser sciatic notch

Ischial tuberosity

Ischial ramus

Iliac crest

Anterior superior iliac spine

Anterior inferior iliac spine

Inferior gluteal line

Acetabulum

Arcuate line

Superior ramus of pubis

Pubic crest

Inferior ramus of pubis

Obturator foramen

Ischial ramus

Pubic symphysis (symphyseal surface)

(a) right hip (coxal) bone, lateral view

(b) right hip (coxal) bone, medial view

Bones of the male pelvis.

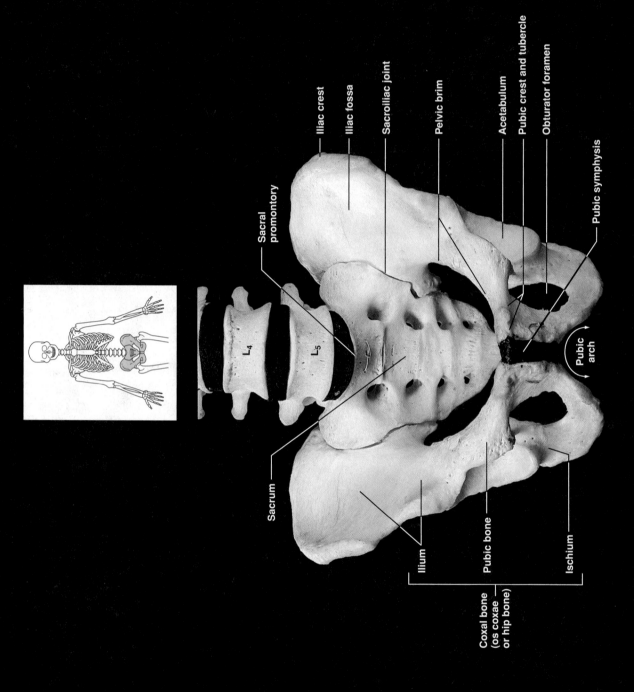

Sacral promontory

Iliac crest

Iliac fossa

Sacroiliac joint

Pelvic brim

Acetabulum

Pubic crest and tubercle

Obturator foramen

Pubic symphysis

Sacrum

Ilium

Pubic bone

Ischium

Coxal bone (os coxae or hip bone)

Pubic arch

L₄

L₅

(c) articulated male pelvis, anterior view

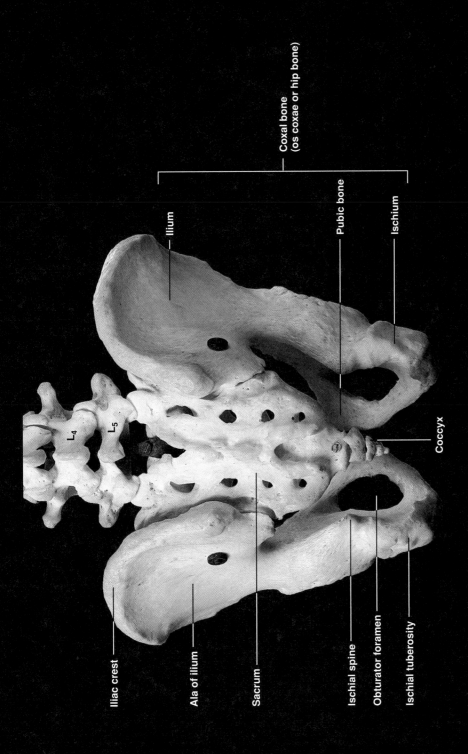

Coxal bone
(os coxae or hip bone)

Ilium

Pubic bone

Ischium

Iliac crest

Ala of ilium

Sacrum

Ischial spine

Obturator foramen

Ischial tuberosity

Coccyx

L4

L5

(d) articulated male pelvis, posterior view

Bones of the male pelvis (continued).

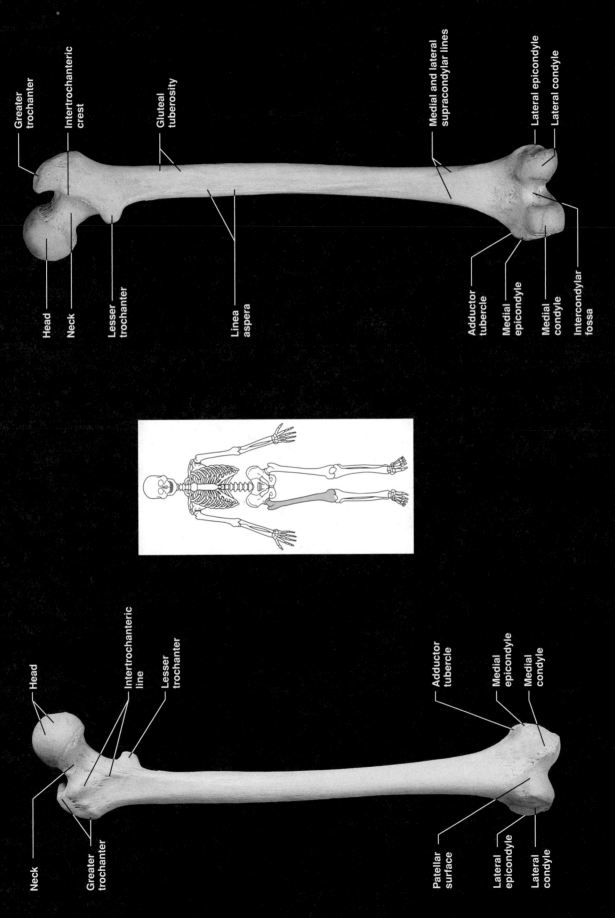

Greater
trochanter

Intertrochanteric
crest

Gluteal
tuberosity

Medial and lateral
supracondylar lines

Lateral epicondyle

Lateral condyle

Head

Neck

Lesser
trochanter

Linea
aspera

Adductor
tubercle

Medial
epicondyle

Medial
condyle

Intercondylar
fossa

(b) posterior view

Head

Intertrochanteric
line

Lesser
trochanter

Neck

Greater
trochanter

Adductor
tubercle

Medial
epicondyle

Medial
condyle

Patellar
surface

Lateral
epicondyle

Lateral
condyle

(a) anterior view

From *A Brief Atlas of the Human Body*, Second Edition. Matt Hutchinson, Jon Mallatt, Elaine N. Marieb, and Patricia Brady Wilhelm. Copyright © 2007 by Pearson Education, Inc. Published by Pearson Benjamin Cummings. All rights reserved.

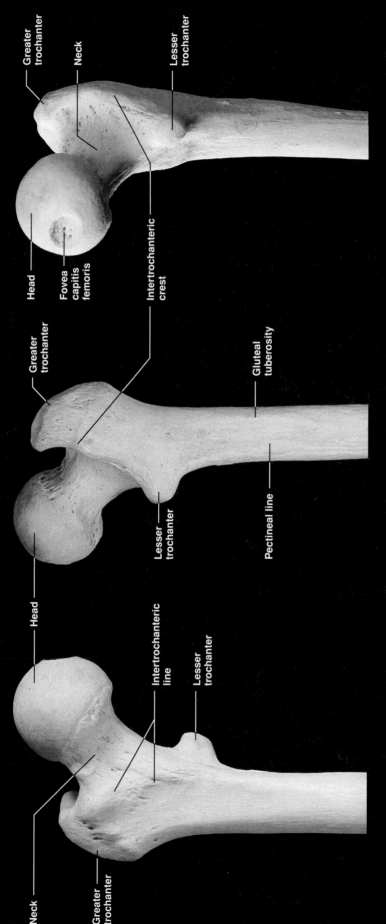

(c) proximal end, anterior view

Neck

Greater
trochanter

Head

Intertrochanteric
line

Lesser
trochanter

(d) proximal end, posterior view

Greater
trochanter

Intertrochanteric
crest

Lesser
trochanter

Pectineal line

Gluteal
tuberosity

(e) proximal end, medial view

Greater
trochanter

Neck

Lesser
trochanter

Head

Fovea
capitis
femoris

Intertrochanteric
crest

Right femur.

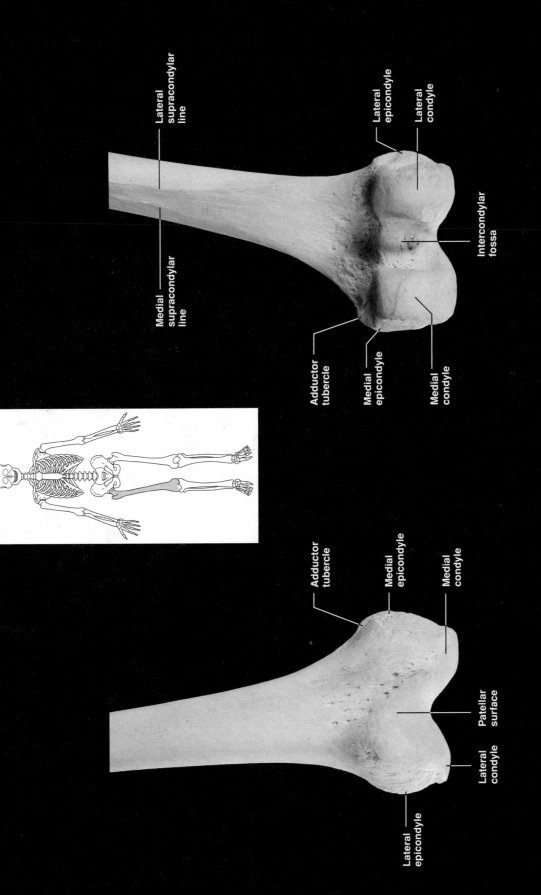

Lateral supracondylar line

Medial supracondylar line

Lateral epicondyle

Lateral condyle

Intercondylar fossa

Adductor tubercle

Medial epicondyle

Medial condyle

(g) distal end, posterior view

Adductor tubercle

Medial epicondyle

Medial condyle

Patellar surface

Lateral condyle

Lateral epicondyle

(f) distal end, anterior view

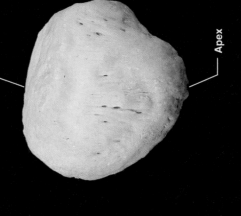

Base

Apex

(i) Right patella, anterior surface

Right femur (continued).

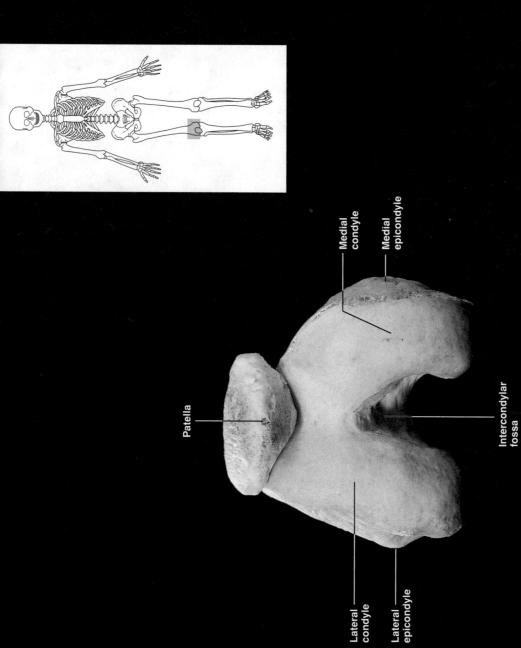

Medial condyle

Medial epicondyle

Patella

Intercondylar fossa

Lateral condyle

Lateral epicondyle

(h) Articulated right femur and patella, inferior view with knee extended

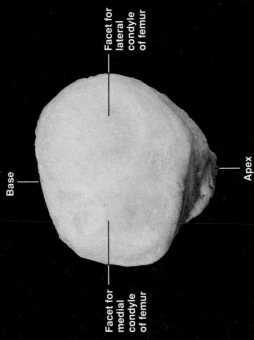

Base

Facet for lateral condyle of femur

Facet for medial condyle of femur

Apex

(k) Right patella, posterior surface

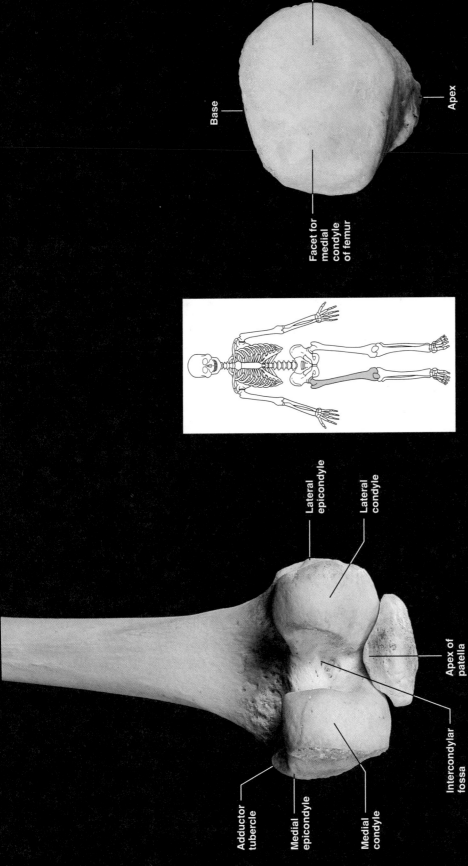

Lateral epicondyle

Lateral condyle

Apex of patella

Intercondylar fossa

Adductor tubercle

Medial epicondyle

Medial condyle

(j) Articulated right femur and patella, inferior posterior view with knee flexed

Right femur (continued).

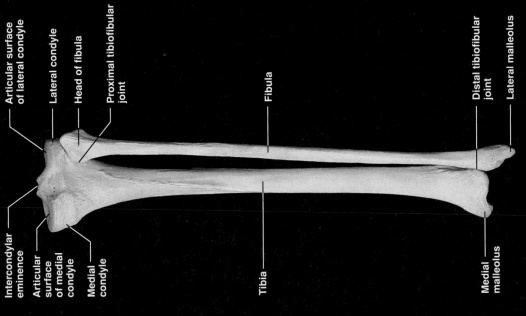

Articular surface
of lateral condyle

Lateral condyle

Head of fibula

Proximal tibiofibular
joint

Intercondylar
eminence

Articular
surface
of medial
condyle

Medial
condyle

Fibula

Tibia

Distal tibiofibular
joint

Lateral malleolus

Medial
malleolus

(b) articulated right tibia and fibula,
posterior view

Right tibia and fibula.

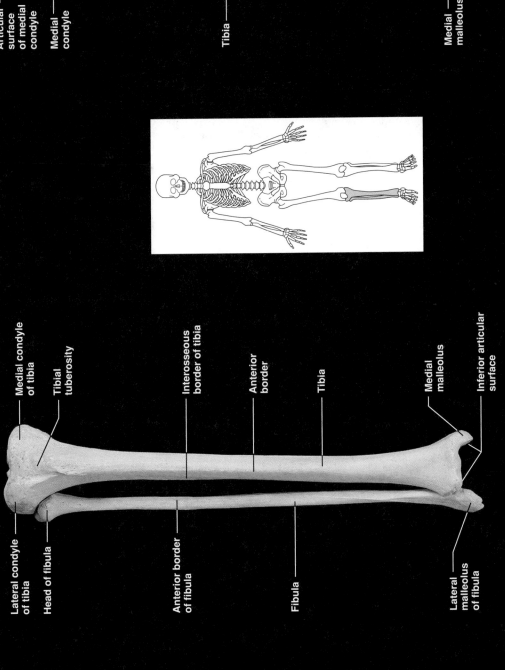

Medial condyle
of tibia

Tibial
tuberosity

Interosseous
border of tibia

Anterior
border

Tibia

Medial
malleolus

Inferior articular
surface

Lateral condyle
of tibia

Head of fibula

Anterior border
of fibula

Fibula

Lateral
malleolus
of fibula

(a) articulated right tibia and fibula,
anterior view

From *A Brief Atlas of the Human Body*, Second Edition. Matt Hutchinson, Jon Mallatt, Elaine N. Marieb, and Patricia Brady Wilhelm. Copyright © 2007 by Pearson Education, Inc.
Published by Pearson Benjamin Cummings. All rights reserved.

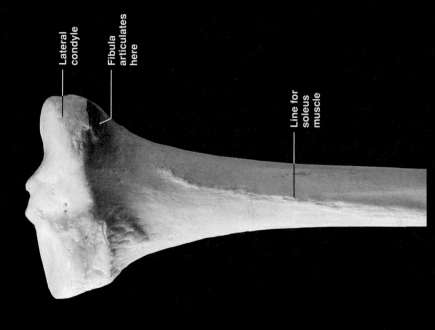

Lateral condyle

Fibula articulates here

Line for soleus muscle

(d) right tibia, proximal end, posterior view

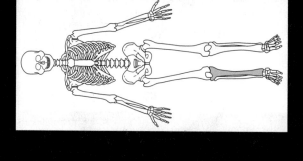

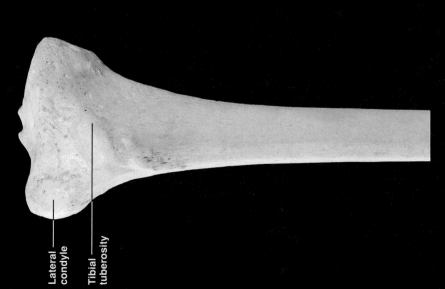

Lateral condyle

Tibial tuberosity

(c) right tibia, proximal end, anterior view

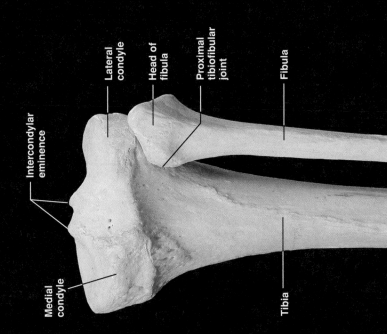

Fibula

Distal tibiofibular joint

Lateral malleolus

Tibia

Medial malleolus

(g) articulated right tibia and fibula, distal end, posterior view

Intercondylar eminence

Lateral condyle

Head of fibula

Proximal tibiofibular joint

Fibula

Medial condyle

Tibia

(f) articulated right tibia and fibula, proximal end, posterior view

Right tibia and fibula (continued)

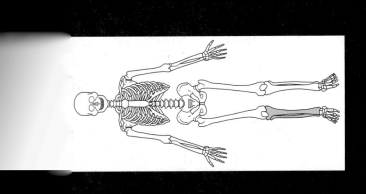

Tibial tuberosity

Anterior intercondylar area

Lateral condyle

Intercondylar eminence

Posterior intercondylar area

Medial condyle

(e) right tibia, proximal end, articular surface

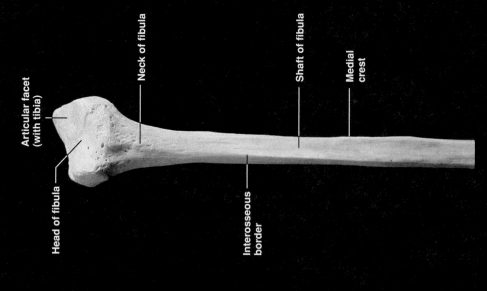

Articular facet (with tibia)

Neck of fibula

Shaft of fibula

Medial crest

Head of fibula

Interosseous border

(j) right fibula, proximal end, medial view

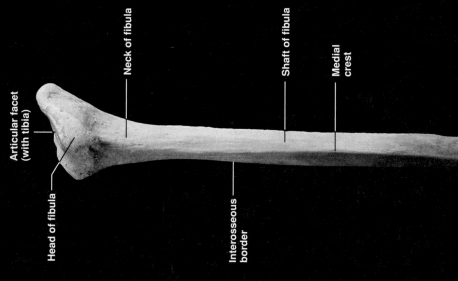

Articular facet (with tibia)

Neck of fibula

Shaft of fibula

Medial crest

Head of fibula

Interosseous border

(i) right fibula, proximal end, posteromedial view

Right tibia and fibula (continued)

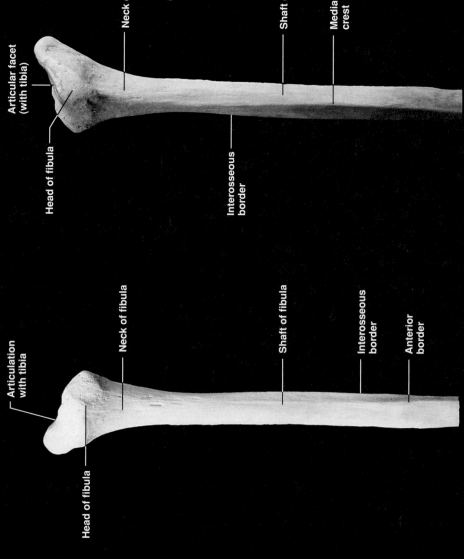

Articulation with tibia

Neck of fibula

Shaft of fibula

Interosseous border

Anterior border

Head of fibula

(h) right fibula, proximal end, anterior view

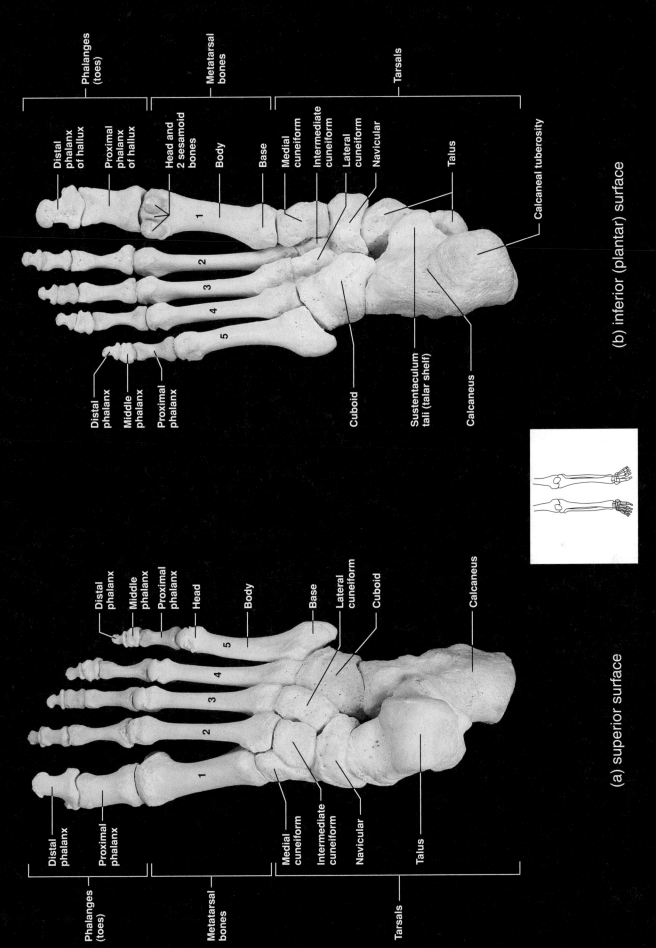

(a) superior surface

(b) inferior (plantar) surface

Bones of the right ankle and foot.

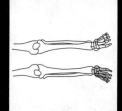

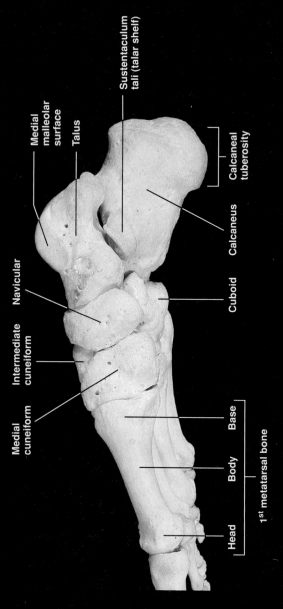

Medial
cuneiform

Intermediate
cuneiform

Navicular

Medial
malleolar
surface

Talus

Sustentaculum
tali (talar shelf)

Head Body Base

1st metatarsal bone

Cuboid Calcaneus

Calcaneal
tuberosity

(c) medial view

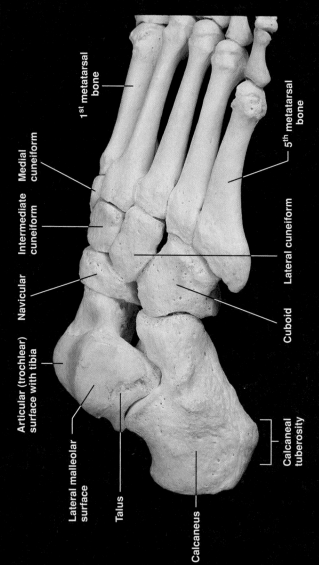

Medial
cuneiform

Intermediate
cuneiform

Navicular

Articular (trochlear)
surface with tibia

Lateral malleolar
surface

Talus

Calcaneus

Calcaneal
tuberosity

Cuboid

Lateral cuneiform

5th metatarsal
bone

1st metatarsal
bone

(d) lateral view

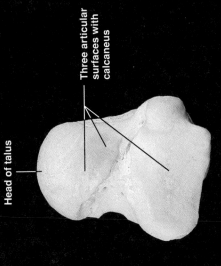

Head of talus

Three articular surfaces with calcaneus

(g) right talus, inferior view

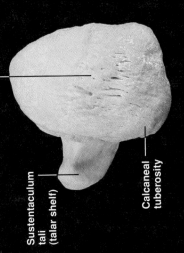

Posterior surface for calcaneal (Achilles) tendon

Sustentaculum tali (talar shelf)

Calcaneal tuberosity

(f) right calcaneus, posterior aspect

Three articular surfaces with talus

Sustentaculum tali (talar shelf)

Posterior surface for calcaneal (Achilles) tendon

(e) right calcaneus, superior aspect

Bones of the right ankle and foot (continued).

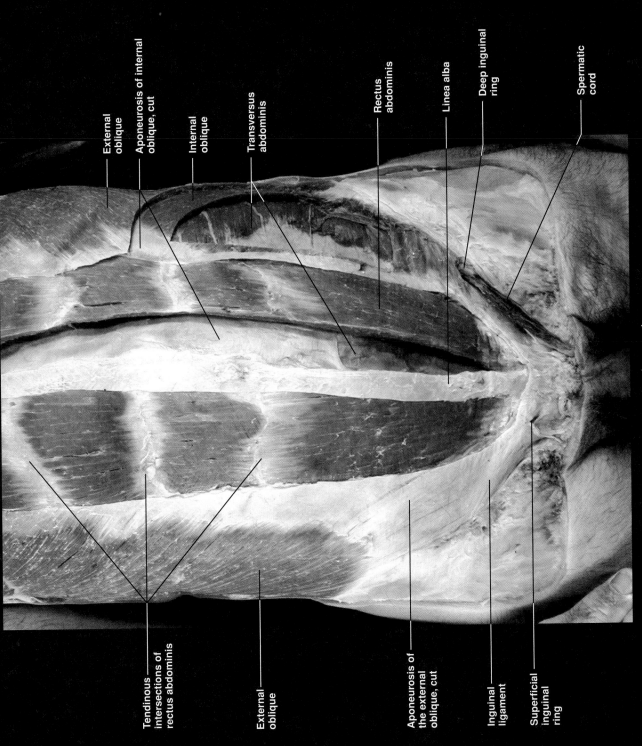

External
oblique

Aponeurosis of internal
oblique, cut

Internal
oblique

Transversus
abdominis

Rectus
abdominis

Linea alba

Deep inguinal
ring

Spermatic
cord

Tendinous
intersections of
rectus abdominis

External
oblique

Aponeurosis of
the external
oblique, cut

Inguinal
ligament

Superficial
inguinal
ring

Abdominal muscles.

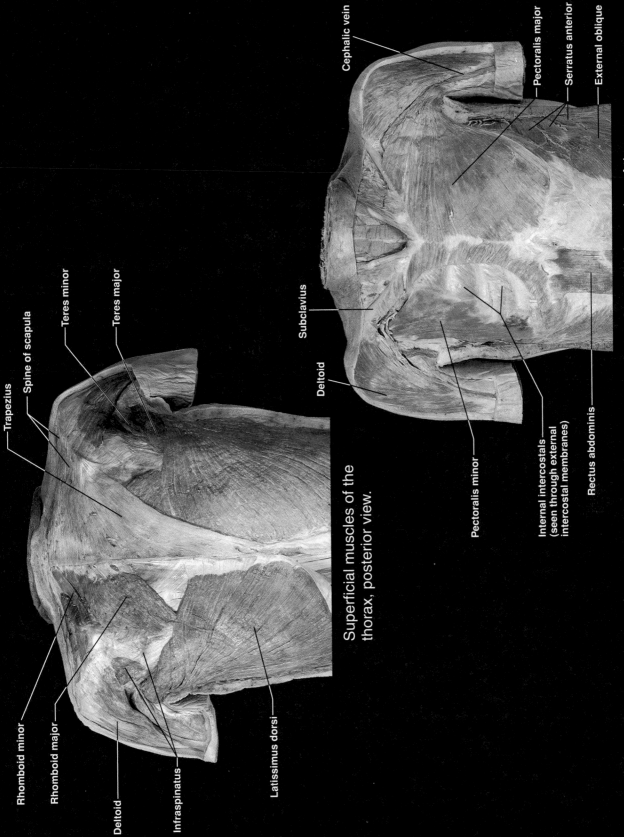

Trapezius

Spine of scapula

Teres minor

Teres major

Rhomboid minor

Rhomboid major

Deltoid

Infraspinatus

Latissimus dorsi

Superficial muscles of the
thorax, posterior view.

Cephalic vein

Pectoralis major

Serratus anterior

External oblique

Subclavius

Deltoid

Pectoralis minor

Internal intercostals
(seen through external
intercostal membranes)

Rectus abdominis

Superficial muscles of the
thorax, anterior view.

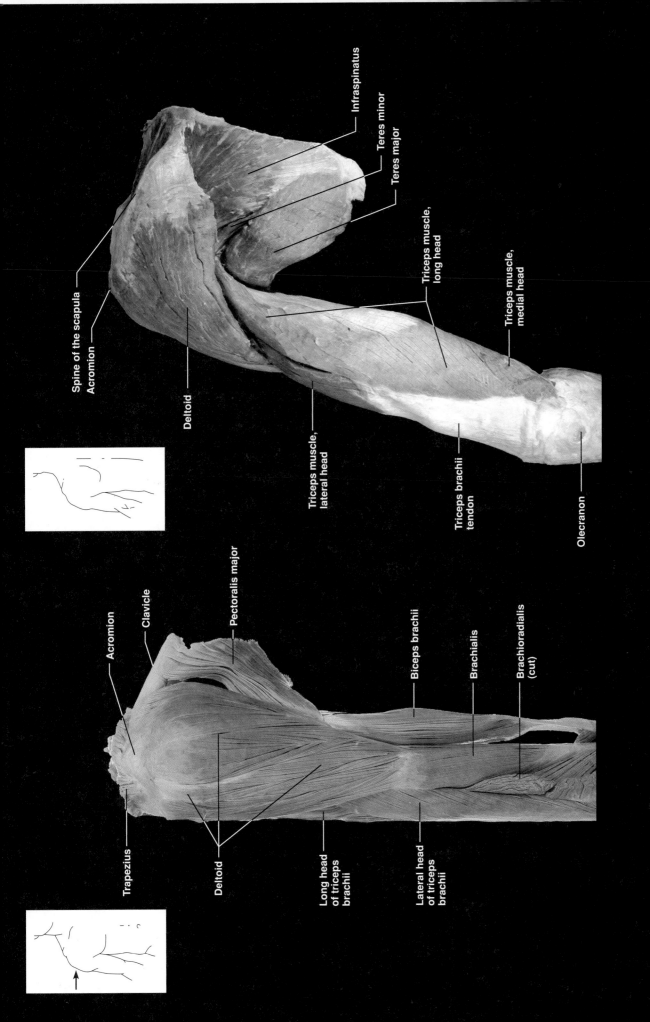

Infraspinatus

Teres minor

Teres major

Triceps muscle, long head

Triceps muscle, medial head

Spine of the scapula

Acromion

Deltoid

Triceps muscle, lateral head

Triceps brachii tendon

Olecranon

Triceps of the left arm, posterior view.

Acromion

Clavicle

Pectoralis major

Trapezius

Deltoid

Long head of triceps brachii

Lateral head of triceps brachii

Biceps brachii

Brachialis

Brachioradialis (cut)

Right shoulder from right, showing deltoid muscle and biceps.

From *A Brief Atlas of the Human Body*, Second Edition. Matt Hutchinson, Jon Mallatt, Elaine N. Marieb, and Patricia Brady Wilhelm. Copyright © 2007 by Pearson Education, Inc. Published by Pearson Benjamin Cummings. All rights reserved.

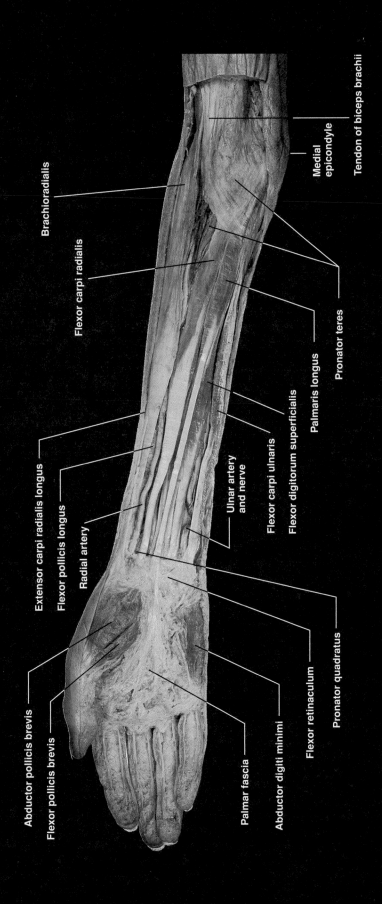

Brachioradialis

Flexor carpi radialis

Extensor carpi radialis longus

Flexor pollicis longus

Radial artery

Abductor pollicis brevis

Flexor pollicis brevis

Palmar fascia

Abductor digiti minimi

Flexor retinaculum

Pronator quadratus

Ulnar artery and nerve

Flexor carpi ulnaris

Flexor digitorum superficialis

Palmaris longus

Pronator teres

Medial epicondyle

Tendon of biceps brachii

(a) palmar surface

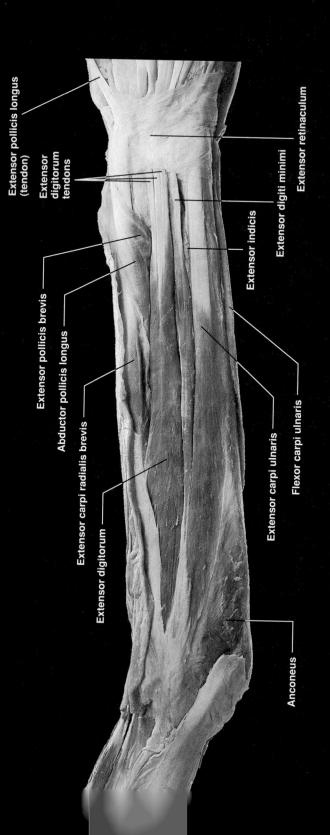

Extensor pollicis longus
(tendon)

Extensor
digitorum
tendons

Extensor retinaculum

Extensor digiti minimi

Extensor indicis

Extensor pollicis brevis

Abductor pollicis longus

Extensor carpi radialis brevis

Extensor carpi ulnaris

Flexor carpi ulnaris

Extensor digitorum

Anconeus

(b) dorsum surface

Right forearm and wrist.

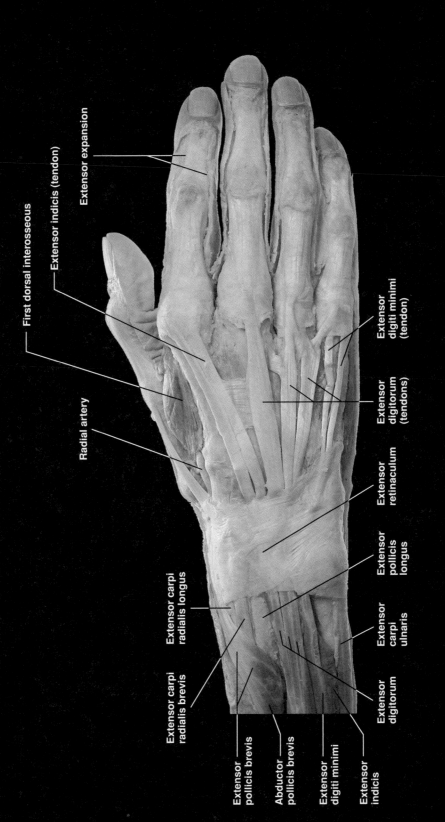

First dorsal interosseous

Extensor indicis (tendon)

Extensor expansion

Radial artery

Extensor digiti minimi (tendon)

Extensor digitorum (tendons)

Extensor retinaculum

Extensor pollicis longus

Extensor carpi ulnaris

Extensor digitorum

Extensor indicis

Extensor digiti minimi

Abductor pollicis brevis

Extensor pollicis brevis

Extensor carpi radialis brevis

Extensor carpi radialis longus

(a) dorsum surface of the right hand and wrist

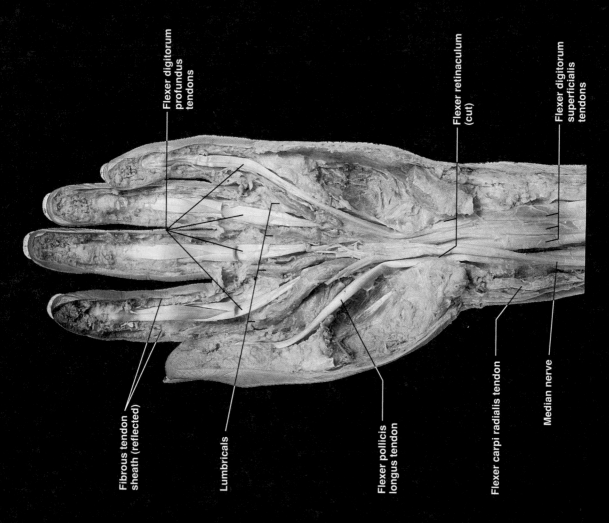

Flexer digitorum profundus tendons

Flexer retinaculum (cut)

Flexer digitorum superficialis tendons

Fibrous tendon sheath (reflected)

Lumbricals

Flexer pollicis longus tendon

Flexer carpi radialis tendon

Median nerve

(b) palmar surface of the left hand and wrist

Wrist and hand.

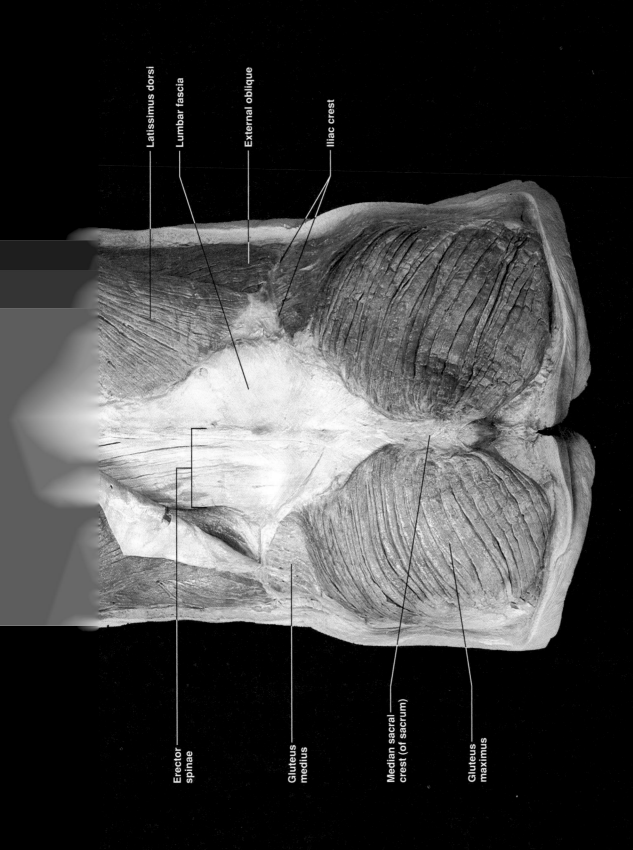

Latissimus dorsi

Lumbar fascia

External oblique

Iliac crest

Erector
spinae

Gluteus
medius

Median sacral
crest (of sacrum)

Gluteus
maximus

Superficial muscles of the superior gluteal region.

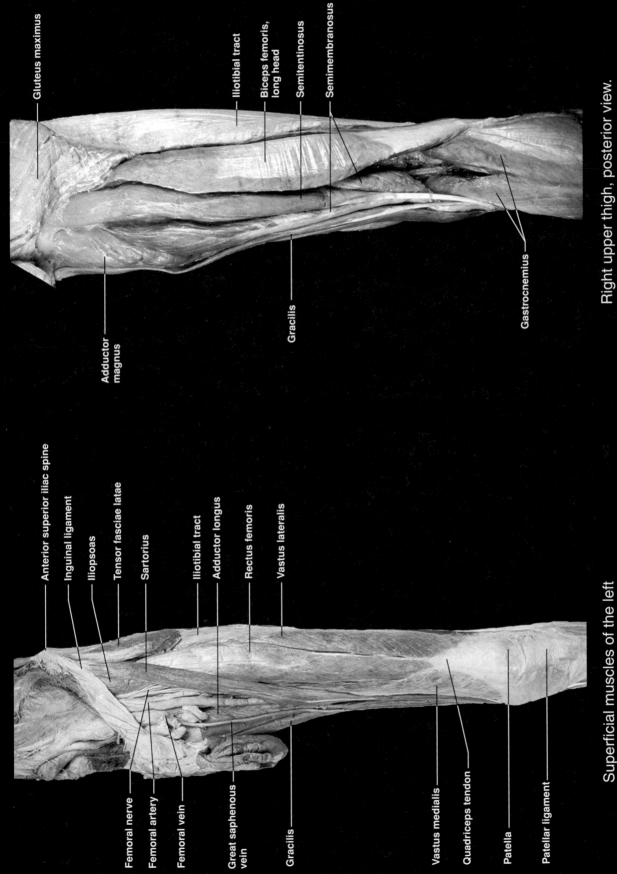

Gluteus maximus

Iliotibial tract

Biceps femoris, long head

Semitentinosus

Semimembranosus

Adductor magnus

Gracilis

Gastrocnemius

Right upper thigh, posterior view.

Anterior superior iliac spine

Inguinal ligament

Iliopsoas

Tensor fasciae latae

Sartorius

Iliotibial tract

Adductor longus

Rectus femoris

Vastus lateralis

Femoral nerve

Femoral artery

Femoral vein

Great saphenous vein

Gracilis

Vastus medialis

Quadriceps tendon

Patella

Patellar ligament

Superficial muscles of the left lower thigh, anterior view.

From *A Brief Atlas of the Human Body*, Second Edition. Matt Hutchinson, Jon Mallatt, Elaine N. Marieb, and Patricia Brady Wilhelm. Copyright © 2007 by Pearson Education, Inc. Published by Pearson Benjamin Cummings. All rights reserved.

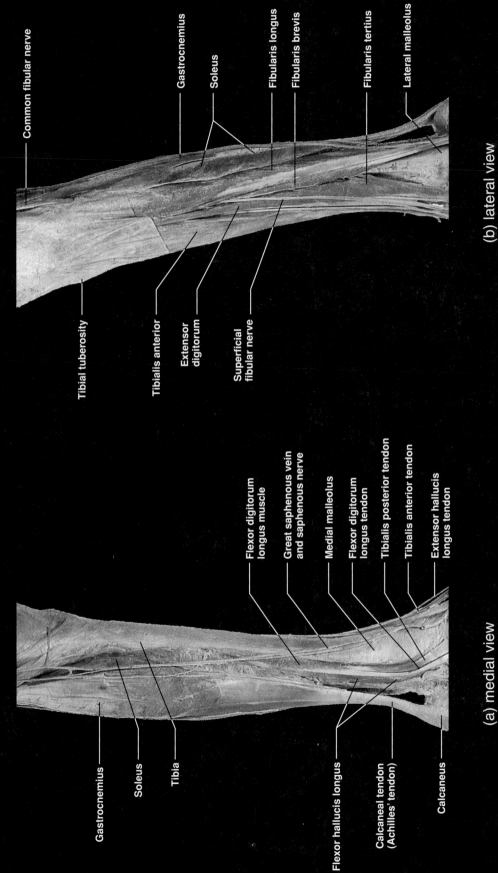

Common fibular nerve

Gastrocnemius

Soleus

Fibularis longus

Fibularis brevis

Fibularis tertius

Lateral malleolus

Tibial tuberosity

Tibialis anterior

Extensor digitorum

Superficial fibular nerve

(b) lateral view

Flexor digitorum longus muscle

Great saphenous vein and saphenous nerve

Medial malleolus

Flexor digitorum longus tendon

Tibialis posterior tendon

Tibialis anterior tendon

Extensor hallucis longus tendon

Gastrocnemius

Soleus

Tibia

Flexor hallucis longus

Calcaneal tendon (Achilles' tendon)

Calcaneus

(a) medial view

Leg.

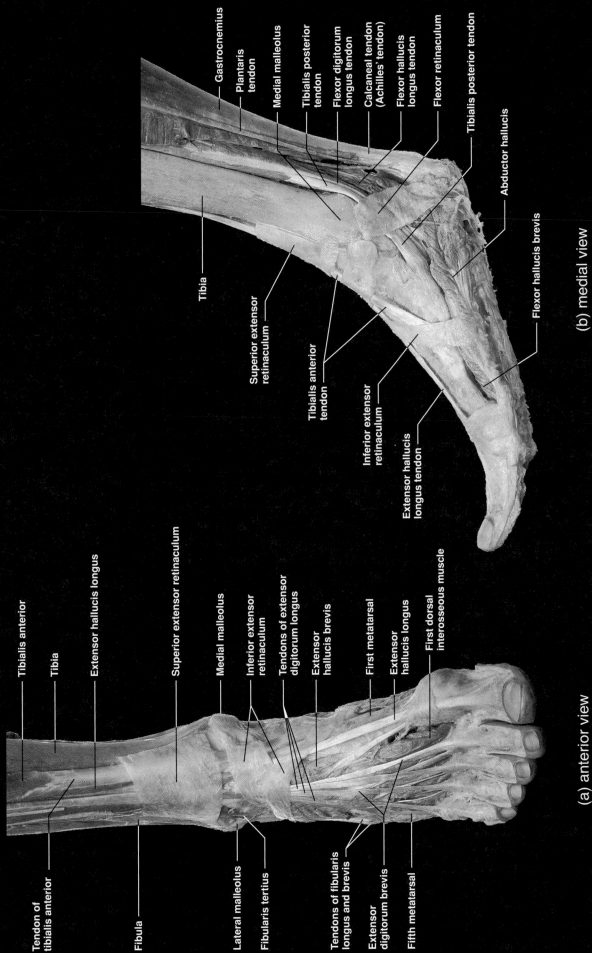

Tendon of
tibialis anterior

Fibula

Tibialis anterior

Tibia

Extensor hallucis longus

Superior extensor retinaculum

Medial malleolus

Inferior extensor
retinaculum

Tendons of extensor
digitorum longus

Extensor
hallucis brevis

First metatarsal

Extensor
hallucis longus

First dorsal
interosseous muscle

Lateral malleolus

Fibularis tertius

Tendons of fibularis
longus and brevis

Extensor
digitorum brevis

Fifth metatarsal

(a) anterior view

Gastrocnemius

Plantaris
tendon

Medial malleolus

Tibialis posterior
tendon

Flexor digitorum
longus tendon

Calcaneal tendon
(Achilles' tendon)

Flexor hallucis
longus tendon

Flexor retinaculum

Tibialis posterior tendon

Abductor hallucis

Flexor hallucis brevis

Tibia

Superior extensor
retinaculum

Tibialis anterior
tendon

Inferior extensor
retinaculum

Extensor hallucis
longus tendon

(b) medial view

Foot.

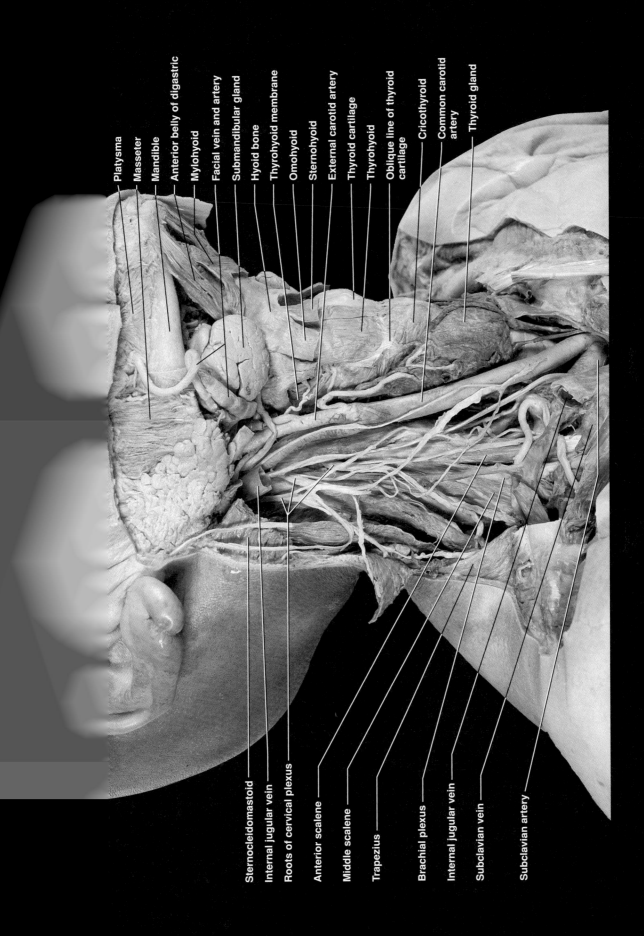

Platysma

Masseter

Mandible

Anterior belly of digastric

Mylohyoid

Facial vein and artery

Submandibular gland

Hyoid bone

Thyrohyoid membrane

Omohyoid

Sternohyoid

External carotid artery

Thyroid cartilage

Thyrohyoid

Oblique line of thyroid cartilage

Cricothyroid

Common carotid artery

Thyroid gland

Sternocleidomastoid

Internal jugular vein

Roots of cervical plexus

Anterior scalene

Middle scalene

Trapezius

Brachial plexus

Internal jugular vein

Subclavian vein

Subclavian artery

Right lower face and upper neck.

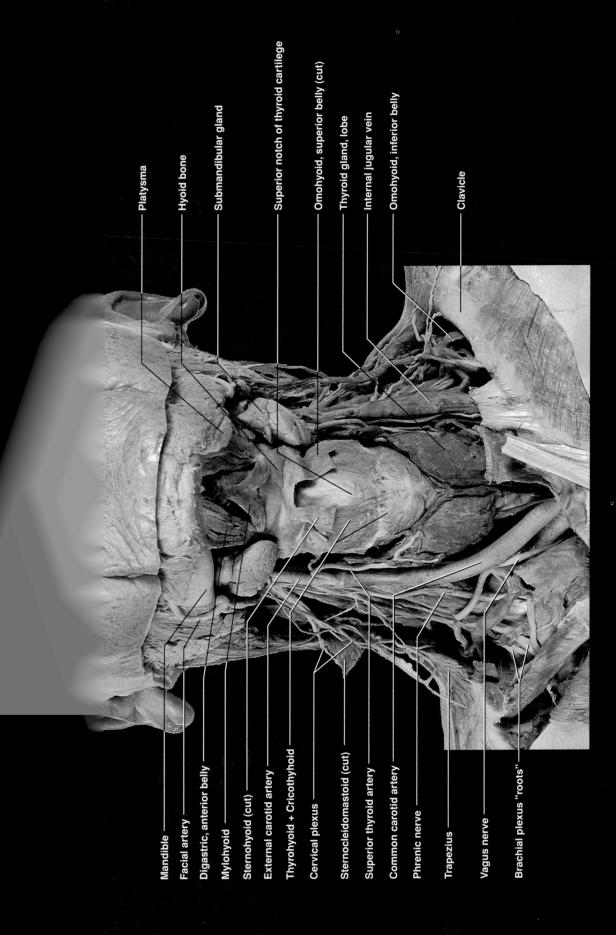

Platysma

Hyoid bone

Submandibular gland

Superior notch of thyroid cartilege

Omohyoid, superior belly (cut)

Thyroid gland, lobe

Internal jugular vein

Omohyoid, inferior belly

Clavicle

Mandible

Facial artery

Digastric, anterior belly

Mylohyoid

Sternohyoid (cut)

External carotid artery

Thyrohyoid + Cricothyroid

Cervical plexus

Sternocleidomastoid (cut)

Superior thyroid artery

Common carotid artery

Phrenic nerve

Trapezius

Vagus nerve

Brachial plexus "roots"

Muscles, blood vessels, and nerves of neck, anterior view.

From *A Brief Atlas of the Human Body*, Second Edition. Matt Hutchinson, Jon Mallatt, Elaine N. Marieb, and Patricia Brady Wilhelm. Copyright © 2007 by Pearson Education, Inc. Published by Pearson Benjamin Cummings. All rights reserved.

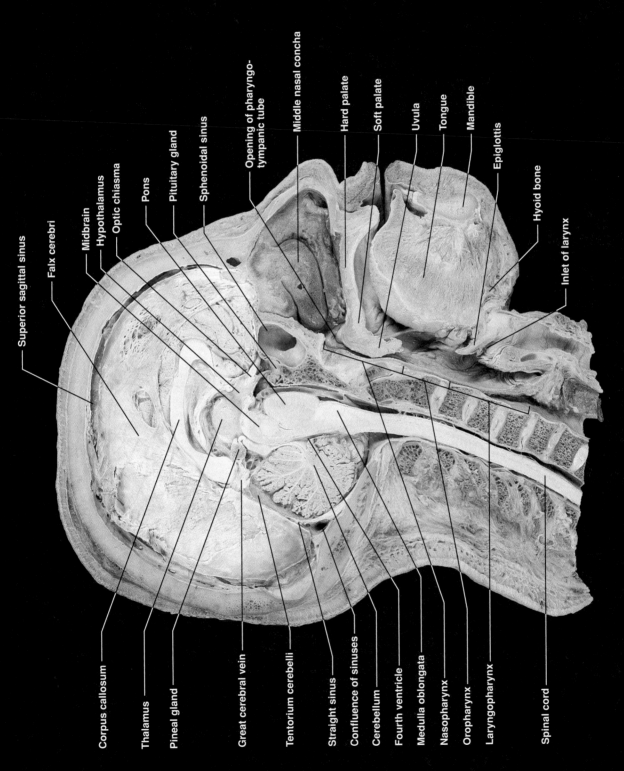

Superior sagittal sinus

Falx cerebri

Midbrain

Hypothalamus

Optic chiasma

Pons

Pituitary gland

Sphenoidal sinus

Opening of pharyngo-
tympanic tube

Middle nasal concha

Hard palate

Soft palate

Uvula

Tongue

Mandible

Epiglottis

Hyoid bone

Inlet of larynx

Corpus callosum

Thalamus

Pineal gland

Great cerebral vein

Tentorium cerebelli

Straight sinus

Confluence of sinuses

Cerebellum

Fourth ventricle

Medulla oblongata

Nasopharynx

Oropharynx

Laryngopharynx

Spinal cord

Sagittal section of the head.

From *A Brief Atlas of the Human Body*, Second Edition. Matt Hutchinson, Jon Mallatt, Elaine N. Marieb, and Patricia Brady Wilhelm. Copyright © 2007 by Pearson Education, Inc. Published by Pearson Benjamin Cummings. All rights reserved.

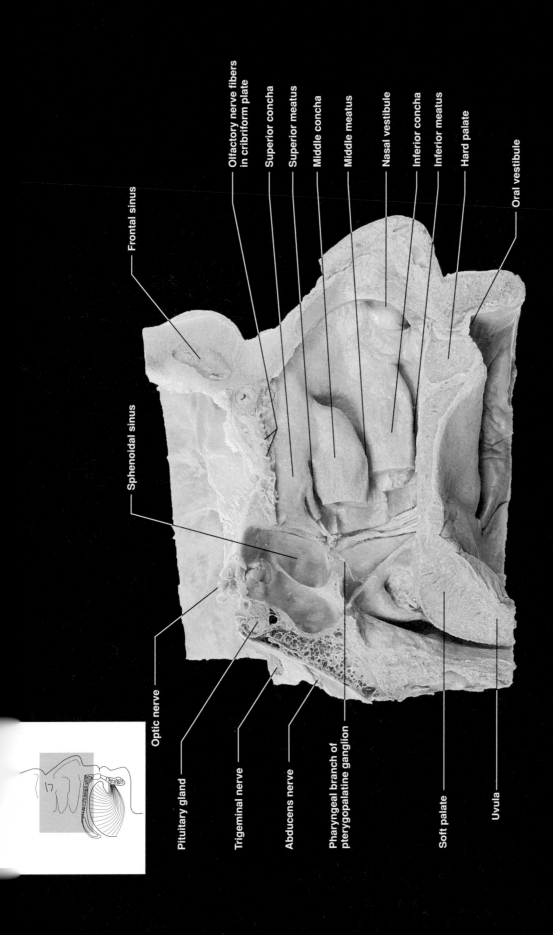

Olfactory nerve fibers
in cribriform plate

Superior concha

Superior meatus

Middle concha

Middle meatus

Nasal vestibule

Inferior concha

Inferior meatus

Hard palate

Oral vestibule

Frontal sinus

Sphenoidal sinus

Optic nerve

Pituitary gland

Trigeminal nerve

Abducens nerve

Pharyngeal branch of
pterygopalatine ganglion

Soft palate

Uvula

Left nasal cavity, lateral wall.

From *A Brief Atlas of the Human Body*, Second Edition. Matt Hutchinson, Jon Mallatt, Elaine N. Marieb, and Patricia Brady Wilhelm. Copyright © 2007 by Pearson Education, Inc.
Published by Pearson Benjamin Cummings. All rights reserved.

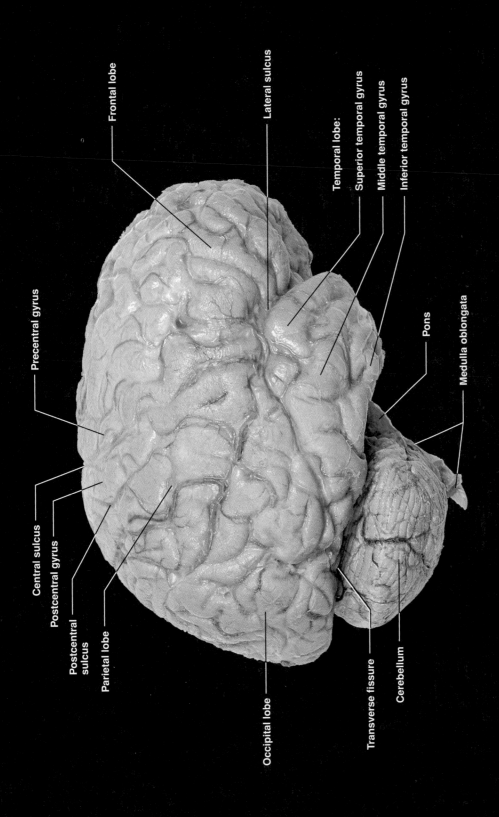

Precentral gyrus

Frontal lobe

Central sulcus

Postcentral gyrus

Lateral sulcus

Postcentral
sulcus

Temporal lobe:

Parietal lobe

Superior temporal gyrus

Middle temporal gyrus

Inferior temporal gyrus

Pons

Medulla oblongata

Occipital lobe

Transverse fissure

Cerebellum

Right cerebral hemisphere (arachnoid mater removed).

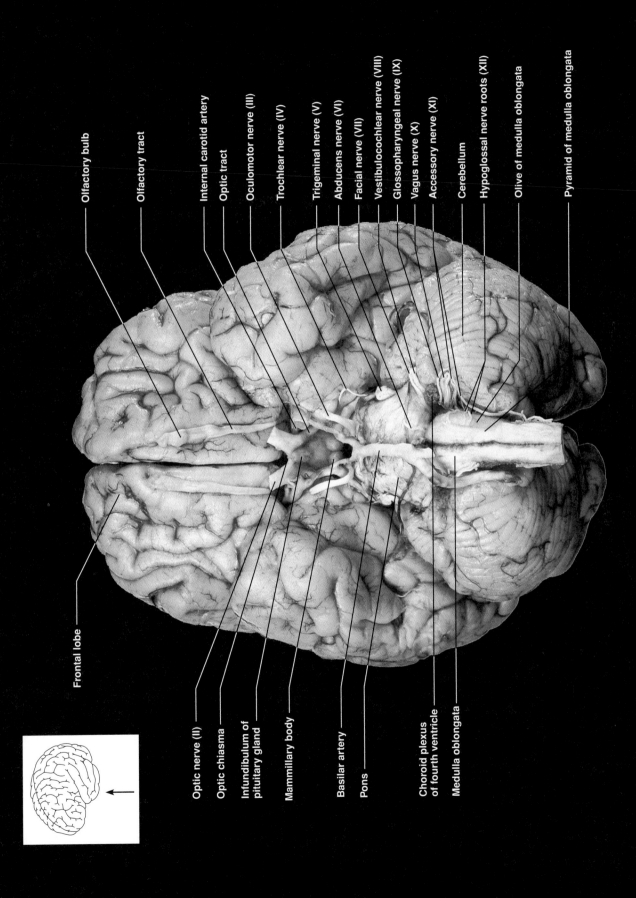

Olfactory bulb

Olfactory tract

Internal carotid artery

Optic tract

Oculomotor nerve (III)

Trochlear nerve (IV)

Trigeminal nerve (V)

Abducens nerve (VI)

Facial nerve (VII)

Vestibulocochlear nerve (VIII)

Glossopharyngeal nerve (IX)

Vagus nerve (X)

Accessory nerve (XI)

Cerebellum

Hypoglossal nerve roots (XII)

Olive of medulla oblongata

Pyramid of medulla oblongata

Frontal lobe

Optic nerve (II)

Optic chiasma

Infundibulum of pituitary gland

Mammillary body

Basilar artery

Pons

Choroid plexus of fourth ventricle

Medulla oblongata

Ventral view of the brain.

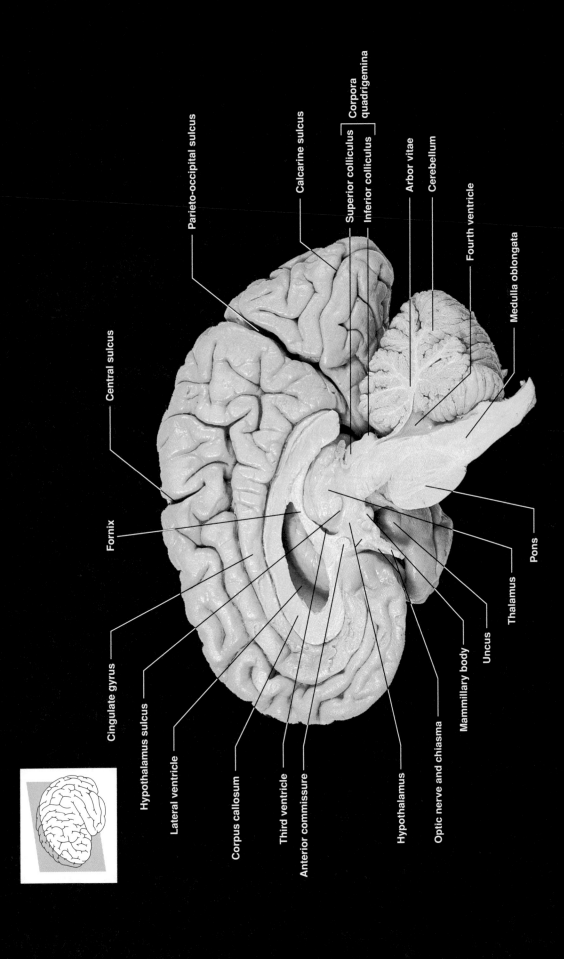

Central sulcus

Cingulate gyrus

Hypothalamus sulcus

Lateral ventricle

Corpus callosum

Third ventricle

Anterior commissure

Fornix

Parieto-occipital sulcus

Calcarine sulcus

Superior colliculus
Inferior colliculus } Corpora quadrigemina

Arbor vitae

Cerebellum

Fourth ventricle

Medulla oblongata

Hypothalamus

Optic nerve and chiasma

Mammillary body

Uncus

Thalamus

Pons

Midsagittal section of the brain.

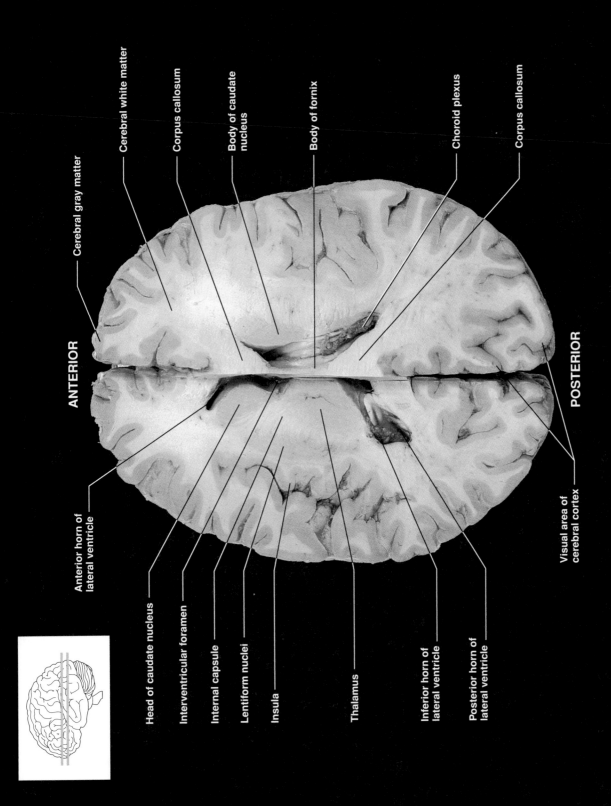

ANTERIOR

Cerebral gray matter

Cerebral white matter

Corpus callosum

Body of caudate nucleus

Body of fornix

Choroid plexus

Corpus callosum

POSTERIOR

Anterior horn of lateral ventricle

Head of caudate nucleus

Interventricular foramen

Internal capsule

Lentiform nuclei

Insula

Thalamus

Inferior horn of lateral ventricle

Posterior horn of lateral ventricle

Visual area of cerebral cortex

Transverse section of the brain, superior view.
Left: on a level with the intraventricular foramen;
right: about 1.5 cm higher.

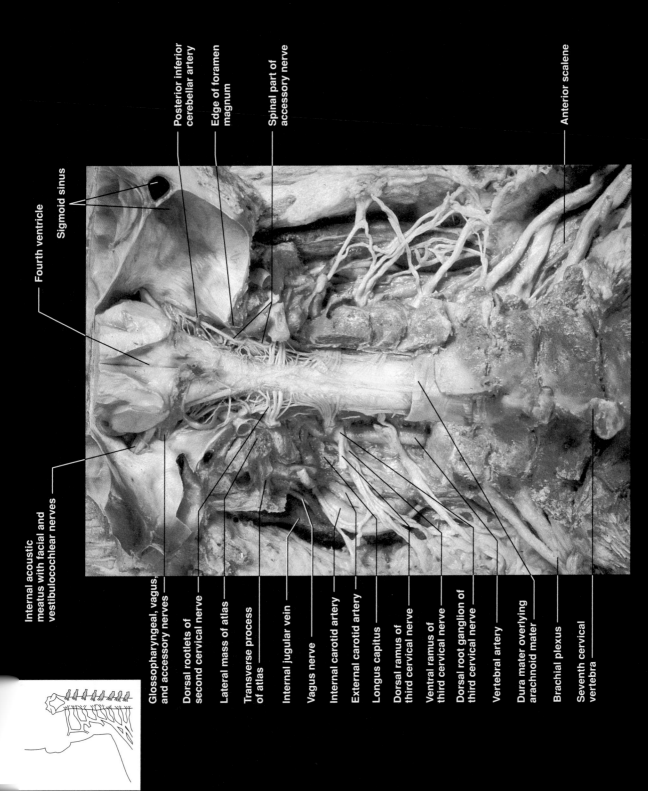

Fourth ventricle

Sigmoid sinus

Posterior inferior cerebellar artery

Edge of foramen magnum

Spinal part of accessory nerve

Anterior scalene

Internal acoustic meatus with facial and vestibulocochlear nerves

Glossopharyngeal, vagus, and accessory nerves

Dorsal rootlets of second cervical nerve

Lateral mass of atlas

Transverse process of atlas

Internal jugular vein

Vagus nerve

Internal carotid artery

External carotid artery

Longus capitus

Dorsal ramus of third cervical nerve

Ventral ramus of third cervical nerve

Dorsal root ganglion of third cervical nerve

Vertebral artery

Dura mater overlying arachnoid mater

Brachial plexus

Seventh cervical vertebra

Brainstem and cervical region of the spinal cord, posterior view.

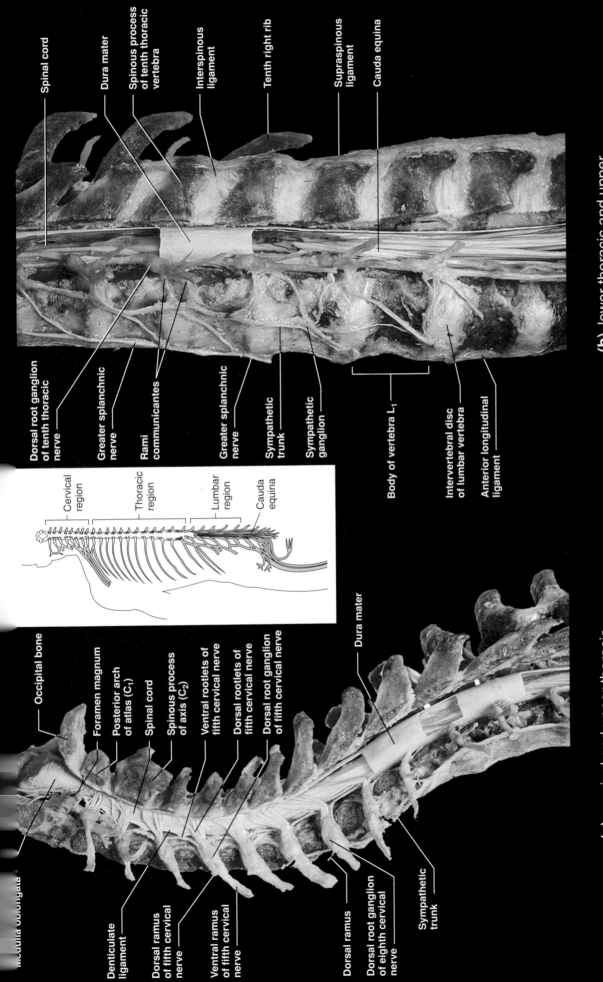

Spinal cord

Dura mater

Spinous process of tenth thoracic vertebra

Interspinous ligament

Tenth right rib

Supraspinous ligament

Cauda equina

Dorsal root ganglion of tenth thoracic nerve

Greater splanchnic nerve

Rami communicantes

Greater splanchnic nerve

Sympathetic trunk

Sympathetic ganglion

Body of vertebra L₁

Intervertebral disc of lumbar vertebra

Anterior longitudinal ligament

(b) lower thoracic and upper lumbar regions from the left

Vertebral column and spinal cord.

Cervical region

Thoracic region

Lumbar region

Cauda equina

Occipital bone

Foramen magnum

Posterior arch of atlas (C₁)

Spinal cord

Spinous process of axis (C₂)

Ventral rootlets of fifth cervical nerve

Dorsal rootlets of fifth cervical nerve

Dorsal root ganglion of fifth cervical nerve

Dura mater

Sympathetic trunk

Medulla oblongata

Denticulate ligament

Dorsal ramus of fifth cervical nerve

Ventral ramus of fifth cervical nerve

Dorsal ramus

Dorsal root ganglion of eighth cervical nerve

Sympathetic trunk

(a) cervical and upper thoracic regions from the left

From *A Brief Atlas of the Human Body*, Second Edition. Matt Hutchinson, Jon Mallatt, Elaine N. Marieb, and Patricia A Brady Wilhelm. Copyright © 2007 by Pearson Education, Inc. Published by Pearson Benjamin Cummings. All rights reserved.

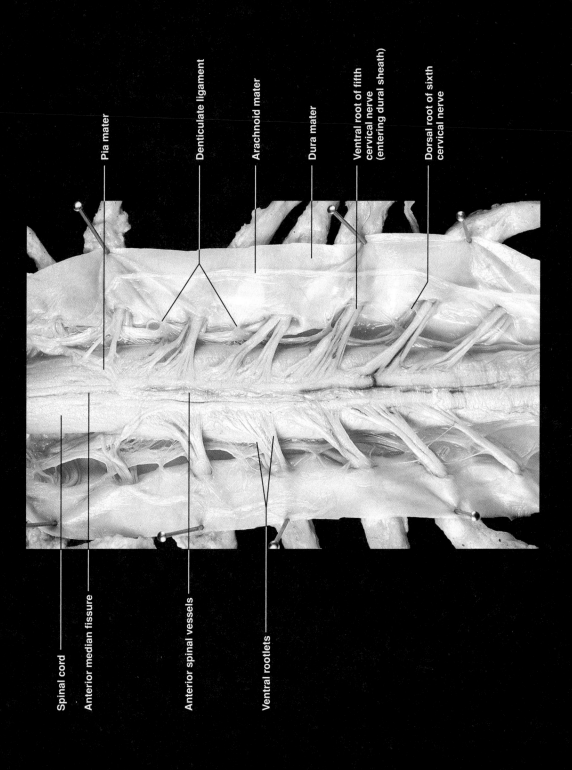

Pia mater

Denticulate ligament

Arachnoid mater

Dura mater

Ventral root of fifth
cervical nerve
(entering dural sheath)

Dorsal root of sixth
cervical nerve

Spinal cord

Anterior median fissure

Anterior spinal vessels

Ventral rootlets

Cervical region of spinal cord, ventral view.

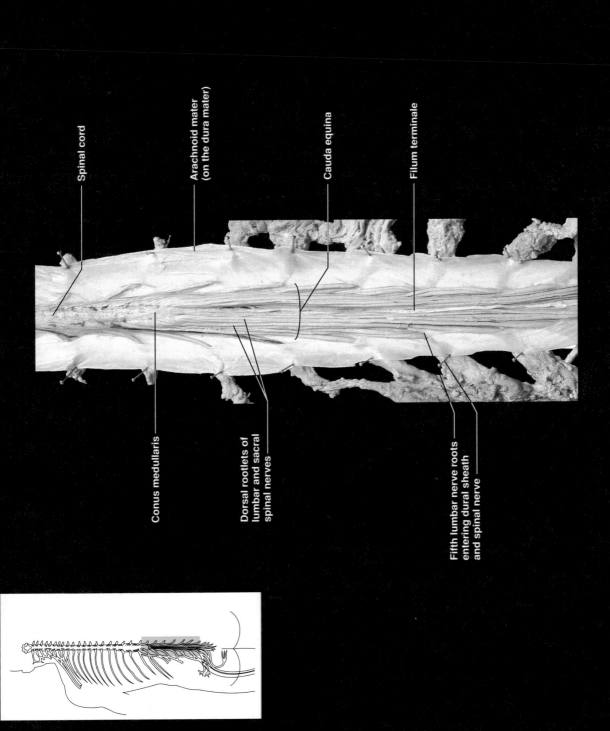

Spinal cord

Arachnoid mater
(on the dura mater)

Cauda equina

Filum terminale

Conus medullaris

Dorsal rootlets of
lumbar and sacral
spinal nerves

Fifth lumbar nerve roots
entering dural sheath
and spinal nerve

Spinal cord and cauda equina, dorsal view of lower end.

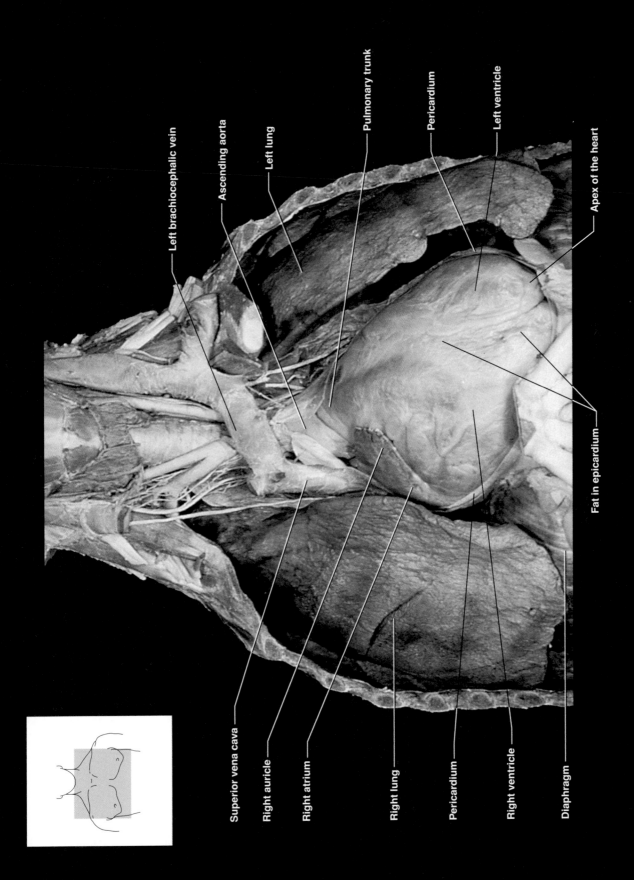

Left brachiocephalic vein

Ascending aorta

Left lung

Pulmonary trunk

Pericardium

Left ventricle

Apex of the heart

Fat in epicardium

Superior vena cava

Right auricle

Right atrium

Right lung

Pericardium

Right ventricle

Diaphragm

Heart and associated structures in thorax.

Ascending aorta

Pulmonary trunk

Auricle of left atrium

Fibrous pericardium

Anterior interventricular branch of left coronary artery

Great cardiac vein

Left ventricle

Visceral layer of serous pericardium (on heart surface)

Right ventricle

Parietal layer of serous pericardium

Superior vena cava

Auricle of right atrium

Right atrium

Anterior cardiac vein

Right coronary artery

Marginal branch of right coronary artery

Diaphragm (covered with parietal pleura)

Heart and pericardium, anterior view.

From *A Brief Atlas of the Human Body*, Second Edition. Matt Hutchinson, Jon Mallatt, Elaine N. Marieb, and Patricia Brady Wilhelm. Copyright © 2007 by Pearson Education, Inc. Published by Pearson Benjamin Cummings. All rights reserved.

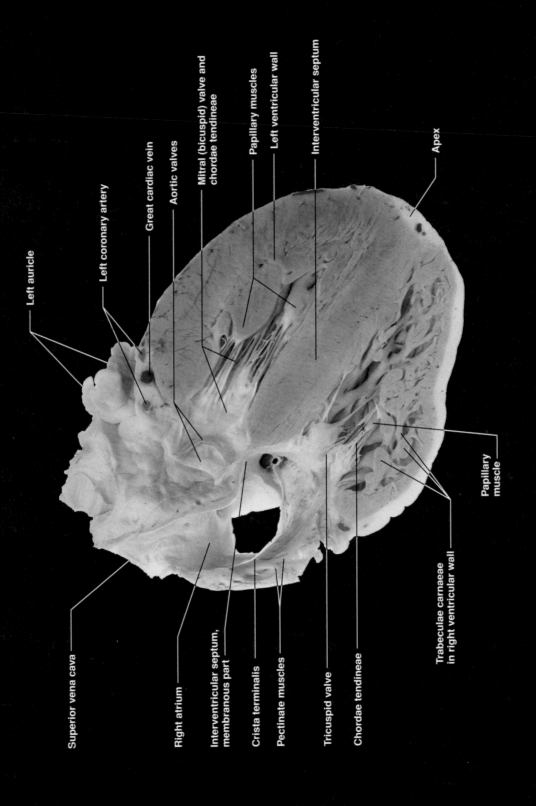

Left auricle

Left coronary artery

Great cardiac vein

Aortic valves

Mitral (bicuspid) valve and
chordae tendineae

Papillary muscles

Left ventricular wall

Interventricular septum

Apex

Superior vena cava

Right atrium

Interventricular septum,
membranous part

Crista terminalis

Pectinate muscles

Tricuspid valve

Chordae tendineae

Trabeculae carnaeae
in right ventricular wall

Papillary
muscle

Coronal section of the ventricles, anterior view.

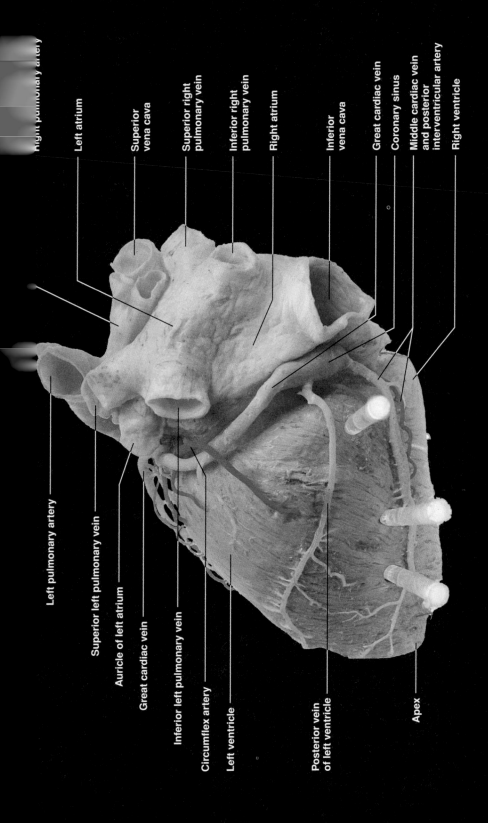

Right pulmonary artery

Left atrium

Superior
vena cava

Superior right
pulmonary vein

Inferior right
pulmonary vein

Right atrium

Inferior
vena cava

Great cardiac vein

Coronary sinus

Middle cardiac vein
and posterior
interventricular artery

Right ventricle

Left pulmonary artery

Superior left pulmonary vein

Auricle of left atrium

Great cardiac vein

Inferior left pulmonary vein

Circumflex artery

Left ventricle

Posterior vein
of left ventricle

Apex

Heart, posterior view (blood vessels injected).

From *A Brief Atlas of the Human Body*, Second Edition. Matt Hutchinson, Jon Mallatt, Elaine N. Marieb, and Patricia Brady Wilhelm. Copyright © 2007 by Pearson Education, Inc.
Published by Pearson Benjamin Cummings. All rights reserved.

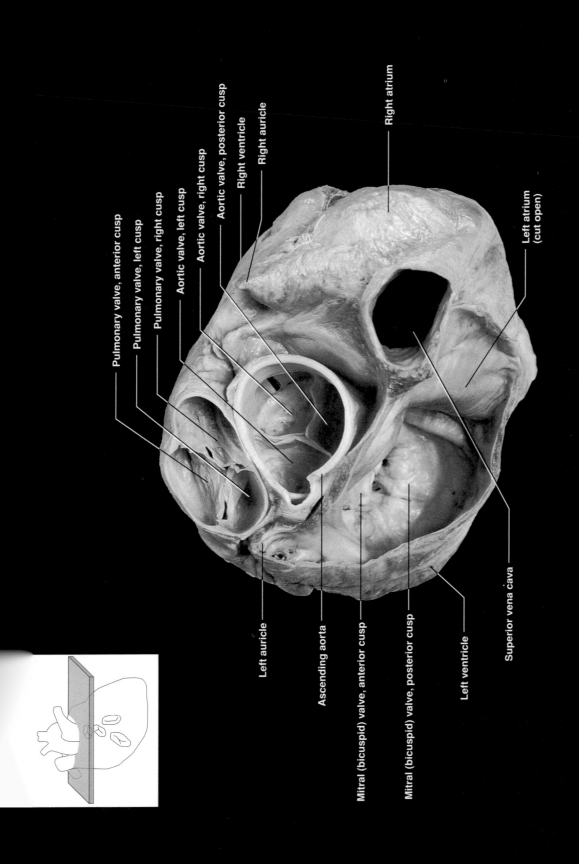

Pulmonary valve, anterior cusp

Pulmonary valve, left cusp

Pulmonary valve, right cusp

Aortic valve, left cusp

Aortic valve, right cusp

Aortic valve, posterior cusp

Right ventricle

Right auricle

Right atrium

Left atrium (cut open)

Left auricle

Ascending aorta

Mitral (bicuspid) valve, anterior cusp

Mitral (bicuspid) valve, posterior cusp

Left ventricle

Superior vena cava

Pulmonary, aortic, and mitral valves of the heart, superior view.

From *A Brief Atlas of the Human Body*, Second Edition. Matt Hutchinson, Jon Mallatt, Elaine N. Marieb, and Patricia Brady Wilhelm. Copyright © 2007 by Pearson Education, Inc. Published by Pearson Benjamin Cummings. All rights reserved.

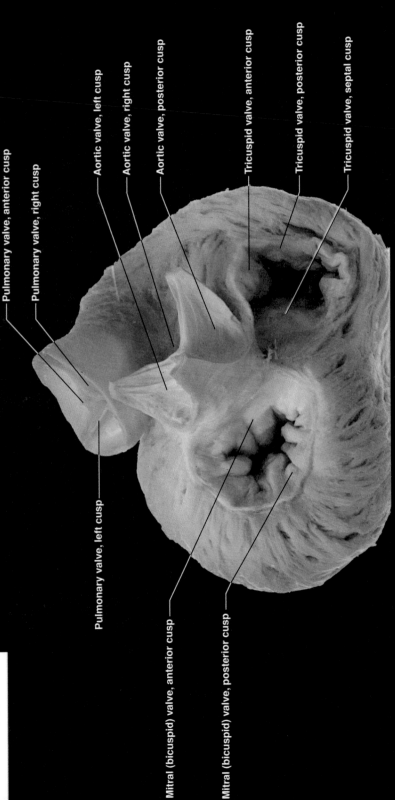

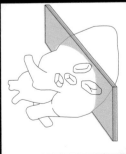

Fibrous framework of the heart (atria removed), posterior view, from the right.

Pulmonary valve, anterior cusp

Pulmonary valve, right cusp

Aortic valve, left cusp

Aortic valve, right cusp

Aortic valve, posterior cusp

Tricuspid valve, anterior cusp

Tricuspid valve, posterior cusp

Tricuspid valve, septal cusp

Pulmonary valve, left cusp

Mitral (bicuspid) valve, anterior cusp

Mitral (bicuspid) valve, posterior cusp

From *A Brief Atlas of the Human Body*, Second Edition. Matt Hutchinson, Jon Mallatt, Elaine N. Marieb, and Patricia Brady Wilhelm. Copyright © 2007 by Pearson Education, Inc. Published by Pearson Benjamin Cummings. All rights reserved.

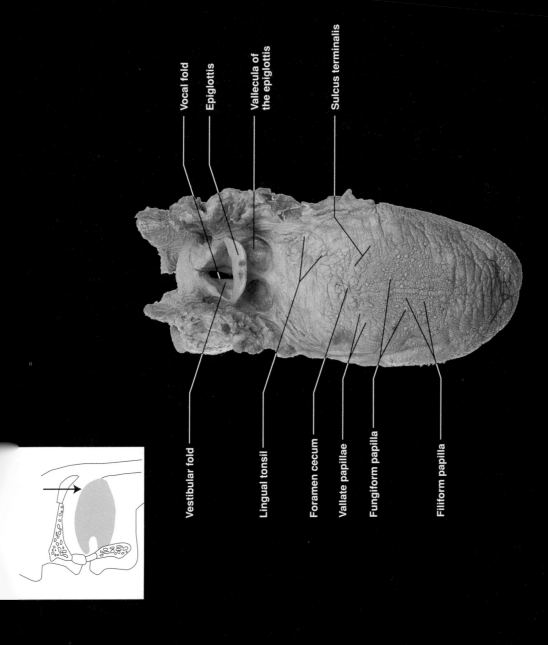

Vocal fold

Epiglottis

Vallecula of
the epiglottis

Sulcus terminalis

Vestibular fold

Lingual tonsil

Foramen cecum

Vallate papillae

Fungiform papilla

Filiform papilla

Tongue and laryngeal inlet.

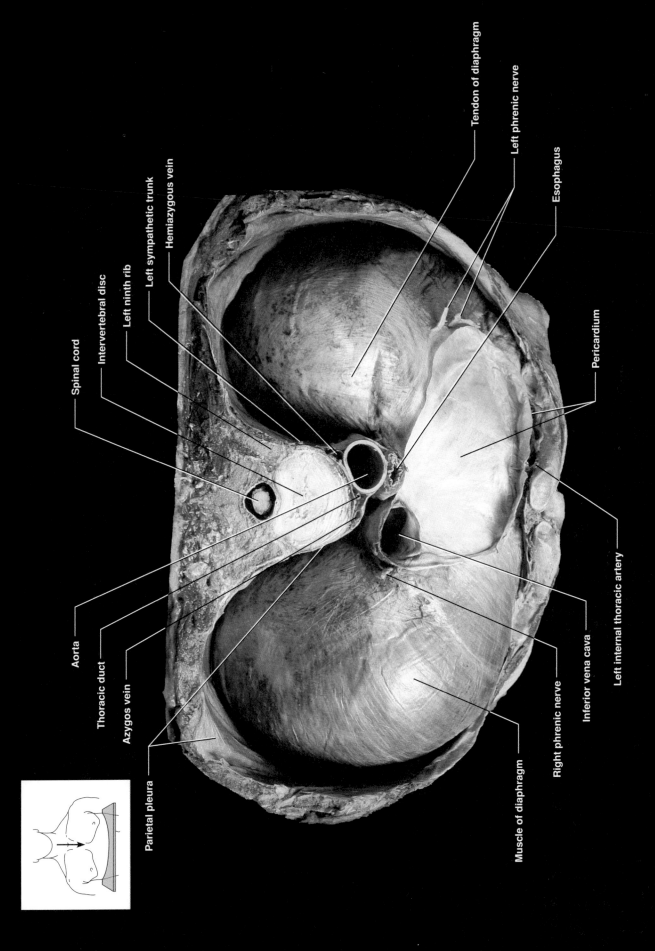

Spinal cord

Intervertebral disc

Left ninth rib

Left sympathetic trunk

Hemiazygous vein

Tendon of diaphragm

Left phrenic nerve

Esophagus

Pericardium

Aorta

Thoracic duct

Azygos vein

Parietal pleura

Muscle of diaphragm

Right phrenic nerve

Inferior vena cava

Left internal thoracic artery

Diaphragm, superior view.

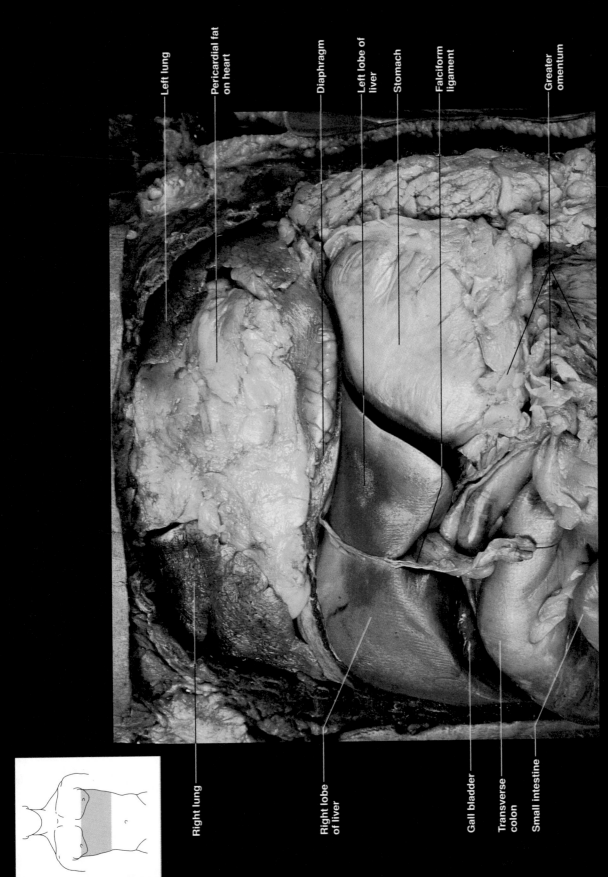

Left lung

Pericardial fat on heart

Diaphragm

Left lobe of liver

Stomach

Falciform ligament

Greater omentum

Right lung

Right lobe of liver

Gall bladder

Transverse colon

Small intestine

(a) upper abdominal viscera, anterior view

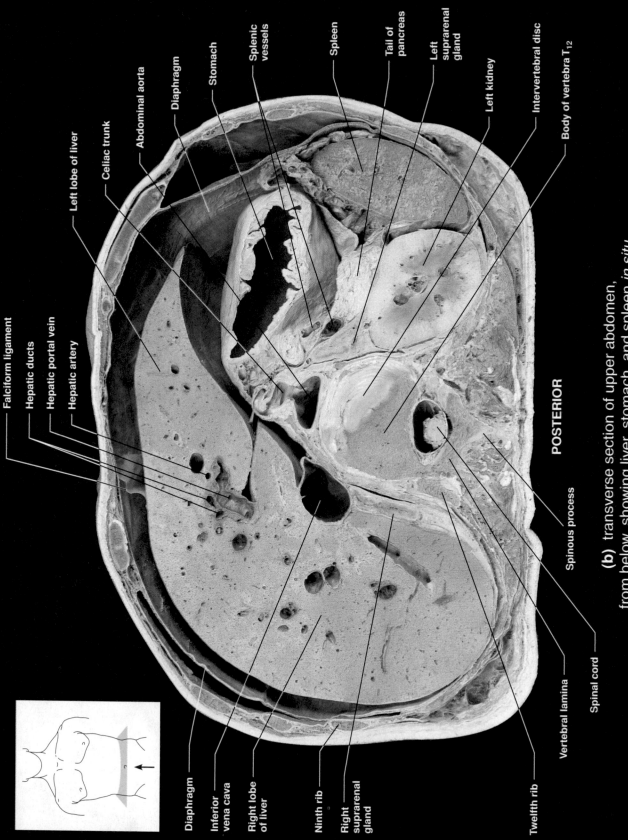

ANTERIOR

Falciform ligament
Hepatic ducts
Hepatic portal vein
Hepatic artery

Left lobe of liver
Celiac trunk
Abdominal aorta
Diaphragm
Stomach
Splenic vessels
Spleen
Tail of pancreas
Left suprarenal gland
Left kidney
Intervertebral disc
Body of vertebra T$_{12}$

POSTERIOR

Diaphragm
Inferior vena cava
Right lobe of liver
Ninth rib
Right suprarenal gland
Twelfth rib
Vertebral lamina
Spinal cord
Spinous process

(b) transverse section of upper abdomen, from below, showing liver, stomach, and spleen *in situ*

Upper abdomen.

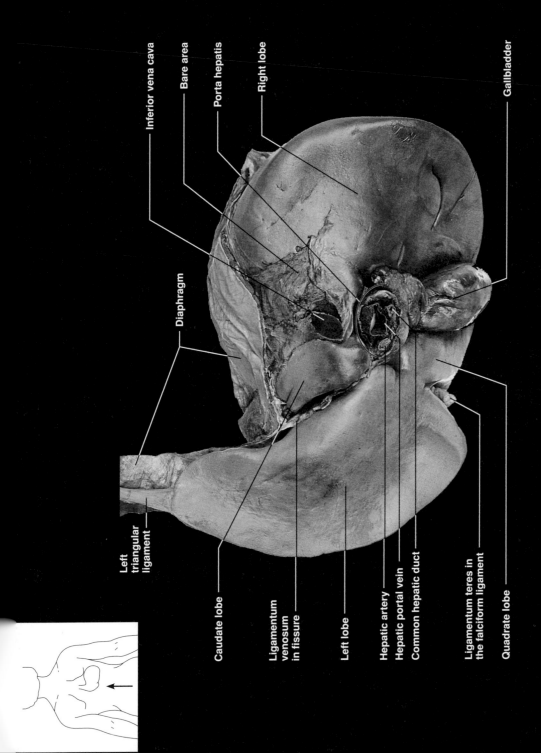

Inferior vena cava

Bare area

Porta hepatis

Right lobe

Gallbladder

Diaphragm

Left triangular ligament

Caudate lobe

Ligamentum venosum in fissure

Left lobe

Hepatic artery

Hepatic portal vein

Common hepatic duct

Ligamentum teres in the falciform ligament

Quadrate lobe

Liver, posteroinferior view.

From *A Brief Atlas of the Human Body*, Second Edition. Matt Hutchinson, Jon Mallatt, Elaine N. Marieb, and Patricia Brady Wilhelm. Copyright © 2007 by Pearson Education, Inc. Published by Pearson Benjamin Cummings. All rights reserved.

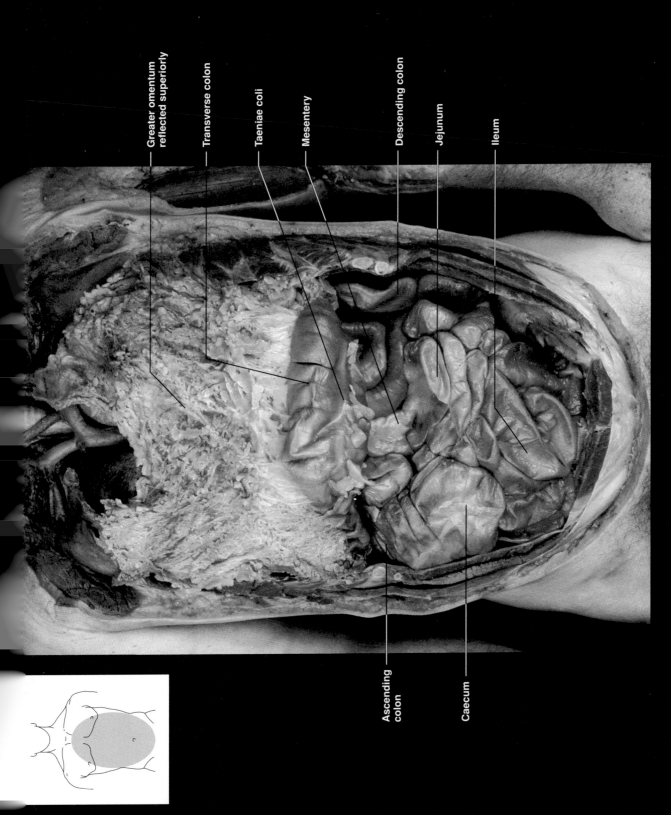

Greater omentum
reflected superiorly

Transverse colon

Taeniae coli

Mesentery

Descending colon

Jejunum

Ileum

Ascending
colon

Caecum

Lower abdominal organs.

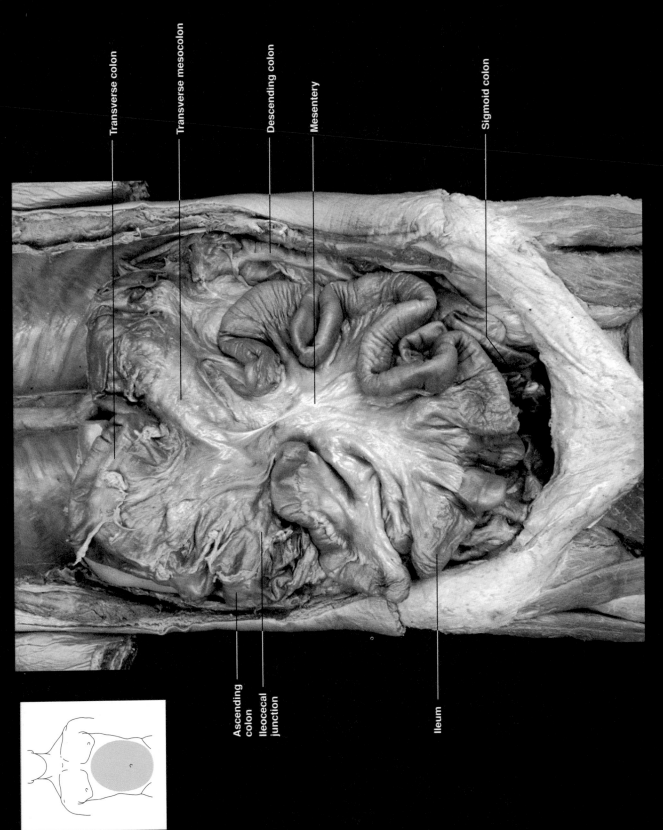

Transverse colon

Transverse mesocolon

Descending colon

Mesentery

Sigmoid colon

Ascending colon

Ileocecal junction

Ileum

Small intestine and colon.

From *A Brief Atlas of the Human Body*, Second Edition. Matt Hutchinson, Jon Mallatt, Elaine N. Marieb, and Patricia Brady Wilhelm. Copyright © 2007 by Pearson Education, Inc. Published by Pearson Benjamin Cummings. All rights reserved.

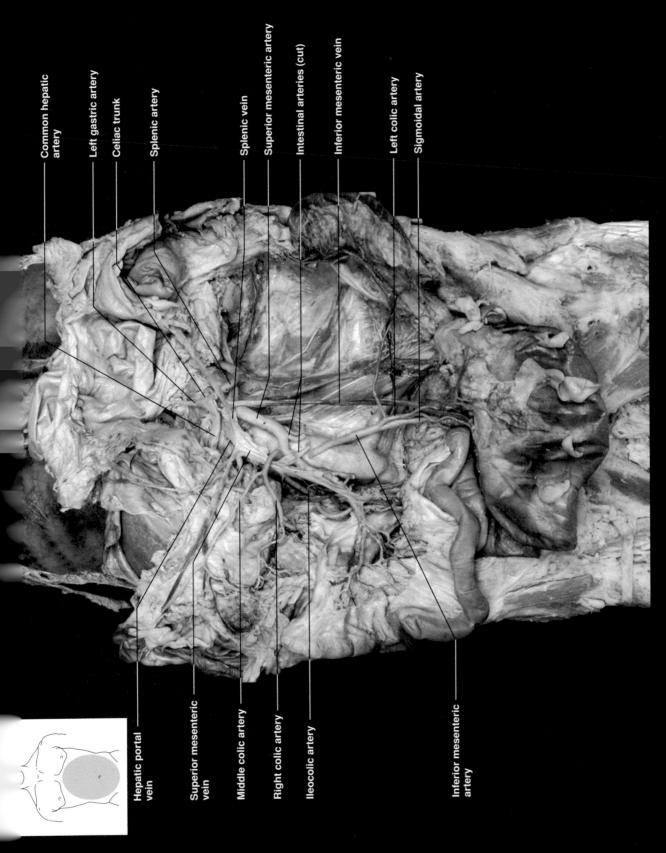

Common hepatic
artery

Left gastric artery

Celiac trunk

Splenic artery

Splenic vein

Superior mesenteric artery

Intestinal arteries (cut)

Inferior mesenteric vein

Left colic artery

Sigmoidal artery

Hepatic portal
vein

Superior mesenteric
vein

Middle colic artery

Right colic artery

Ileocolic artery

Inferior mesenteric
artery

Vessels of the gastrointestinal organs.

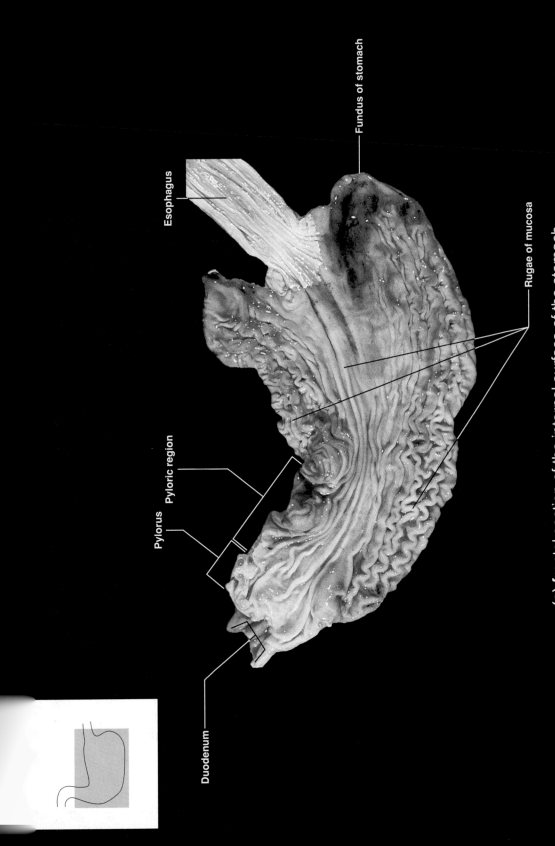

Esophagus

Fundus of stomach

Rugae of mucosa

Pyloric region

Pylorus

Duodenum

(a) frontal section of the internal surface of the stomach.

From *A Brief Atlas of the Human Body*, Second Edition. Matt Hutchinson, Jon Mallatt, Elaine N. Marieb, and Patricia Brady Wilhelm. Copyright © 2007 by Pearson Education, Inc. Published by Pearson Benjamin Cummings. All rights reserved.

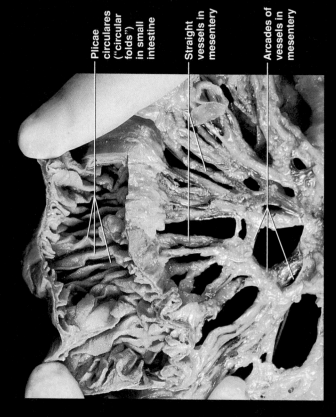

Plicae
circulares
("circular
folds")
in small
intestine

Straight
vessels in
mesentery

Arcades of
vessels in
mesentery

(b) small intestine, cut open to show
plicae circulares

nternal surfaces of the stomach and small intestine

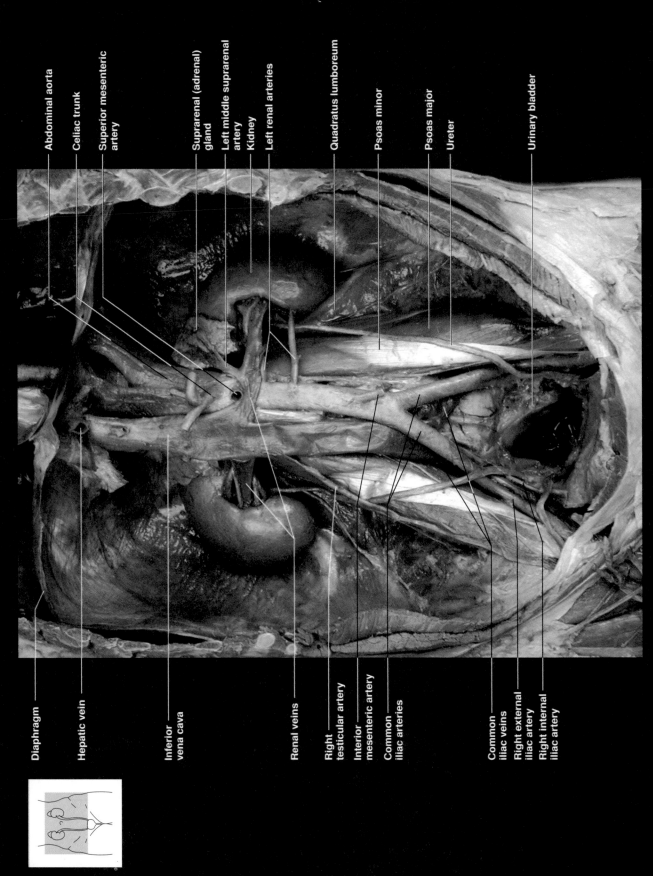

Diaphragm

Hepatic vein

Inferior
vena cava

Renal veins

Right
testicular artery

Interior
mesenteric artery

Common
iliac arteries

Common
iliac veins

Right external
iliac artery

Right internal
iliac artery

Abdominal aorta

Celiac trunk

Superior mesenteric
artery

Suprarenal (adrenal)
gland

Left middle suprarenal
artery

Kidney

Left renal arteries

Quadratus lumboreum

Psoas minor

Psoas major

Ureter

Urinary bladder

Retroperitoneal abdominal structures.

From *A Brief Atlas of the Human Body*, Second Edition. Matt Hutchinson, Jon Mallatt, Elaine N. Marieb, and Patricia Brady Wilhelm. Copyright © 2007 by Pearson Education, Inc.
Published by Pearson Benjamin Cummings. All rights reserved.

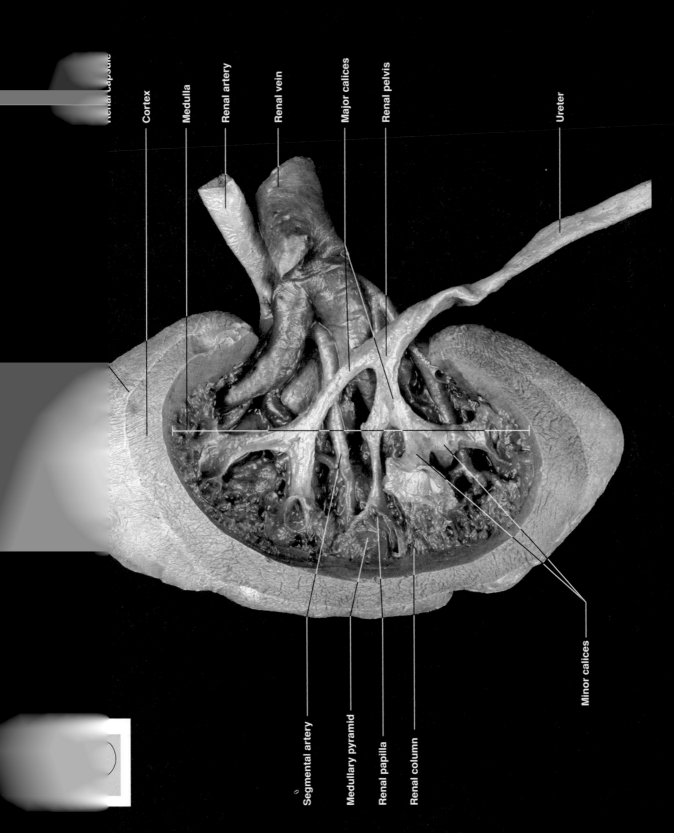

Renal capsule

Cortex

Medulla

Renal artery

Renal vein

Major calices

Renal pelvis

Ureter

Segmental artery

Medullary pyramid

Renal papilla

Renal column

Minor calices

Kidney, internal structure in frontal section

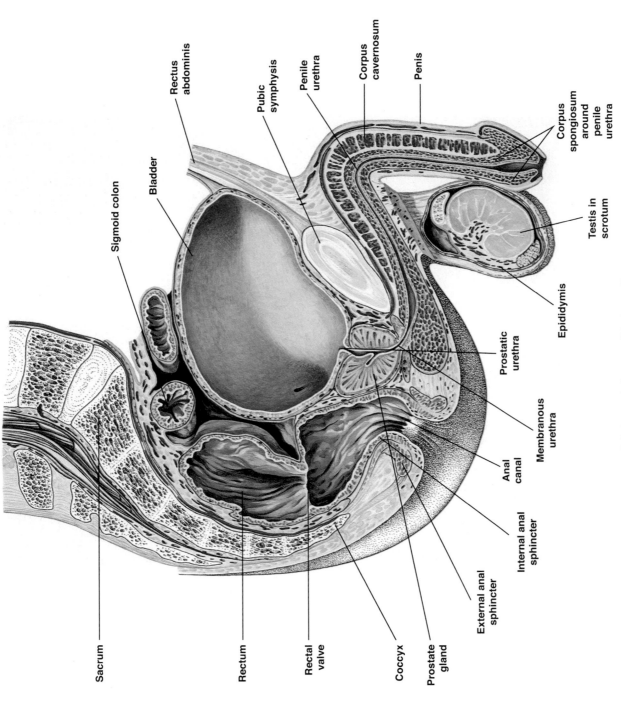

Rectus
abdominis

Bladder

Sigmoid colon

Sacrum

Rectum

Rectal
valve

Coccyx

Prostate
gland

External anal
sphincter

Internal anal
sphincter

Anal
canal

Membranous
urethra

Prostatic
urethra

Epididymis

Testis in
scrotum

Pubic
symphysis

Penile
urethra

Corpus
cavernosum

Penis

Corpus
spongiosum
around
penile
urethra

Male pelvis, sagittal section.
De Agostini/Getty Images

From *A Brief Atlas of the Human Body*, Second Edition. Matt Hutchinson, Jon Mallatt, Elaine N. Marieb, and Patricia Brady Wilhelm. Copyright © 2007 by Pearson Education, Inc.
Published by Pearson Benjamin Cummings. All rights reserved.

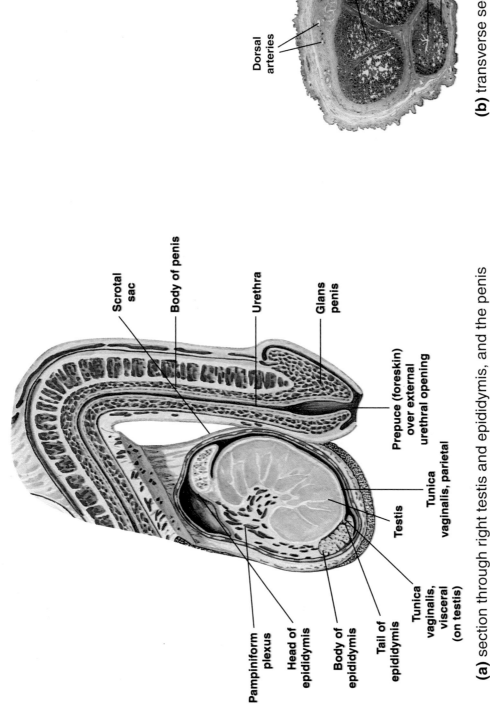

Dorsal arteries

Dorsal vein

Corpus cavernosa

Urethra

Corpus spongiosum

(b) transverse section through penis

Steve Gschmeissner/Science Source

Scrotal sac

Body of penis

Urethra

Glans penis

Pampiniform plexus

Head of epididymis

Body of epididymis

Tail of epididymis

Tunica vaginalis, visceral (on testis)

Testis

Tunica vaginalis, parietal

Prepuce (foreskin) over external urethral opening

(a) section through right testis and epididymis, and the penis

De Agostini/Getty Images

Sections through male reproductive structures.

From *A Brief Atlas of the Human Body*, Second Edition. Matt Hutchinson, Jon Mallatt, Elaine N. Marieb, and Patricia Brady Wilhelm. Copyright © 2007 by Pearson Education, Inc. Published by Pearson Benjamin Cummings. All rights reserved.

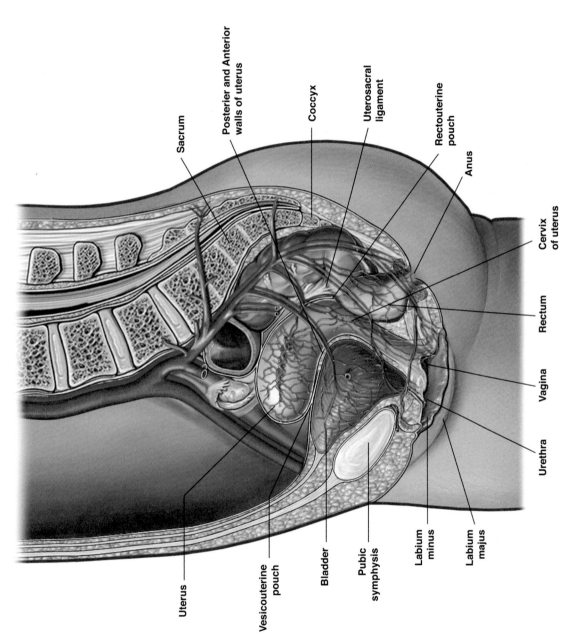

Female pelvis, sagittal section (uterus points forward in this view).

Uterus

Posterier and Anterior walls of uterus

Sacrum

Coccyx

Uterosacral ligament

Rectouterine pouch

Anus

Cervix of uterus

Vesicouterine pouch

Bladder

Pubic symphysis

Rectum

Vagina

Urethra

Labium minus

Labium majus

From *A Brief Atlas of the Human Body*, Second Edition. Matt Hutchinson, Jon Mallatt, Elaine N. Marieb, and Patricia Brady Wilhelm. Copyright © 2007 by Pearson Education, Inc. Published by Pearson Benjamin Cummings. All rights reserved.

Index